FINE PARTICLES IN GASEOUS MEDIA

FINE PARTICLES
IN GASEOUS MEDIA

Howard E. Hesketh, P.E.
Professor of Environmental Engineering
Southern Illinois University
Carbondale, Illinois

ANN ARBOR SCIENCE
PUBLISHERS INC
P.O. BOX 1425 • ANN ARBOR, MICH. 48106

Second Printing, 1979

Copyright © 1977 by Ann Arbor Science Publishers, Inc.
230 Collingwood, P.O. Box 1425, Ann Arbor, Michigan 48106

Library of Congress Catalog Card No. 77-076906
ISBN   0-250-40182-7

22090

PREFACE

The need for knowledge related to the behavior of fine particles increases in importance as the detrimental effects of these materials are reported and control regulations tightened. The *art* of particle control is often not applicable to removal of fine particles, but incorporation of a few basic scientific principles can make the difference. A specific example of forces often overlooked is the positive, rather than the negative application of the phoretic forces. More information in usable form is needed to meet these rising requirements.

In this book, I have tried to compile and organize information relevant to the control of fine particulate matter so that it is presented clearly and concisely and can be easily understood and utilized. This material is intended to be most helpful to practical air pollution control specialists who already have a basic background and who are concerned with the technical aspects of small particles. This work should also be of value to those theoreticians seeking to advance the science of small particle control; however, more theoretical works are available. The material has been gathered from many sources to try to obtain the most up-to-date and useful information on fine particles. Treatment of some areas may be concise, but an effort has been made to serve the needs of the practicing scientist, and no important useful topic is omitted.

Behavior of particulates suspended in a gaseous medium cannot be considered independently of the gas. Appropriate information is provided as necessary throughout the book to account for this gaseous conveying medium. Characteristics of fine particulate matter are presented to convey to the reader that the unexpected becomes more typical as particle size decreases, *i.e.,* very fine particles exhibit unexpected properties. Methods of accounting for variations in size and shape as well as properties are provided, and these must be utilized throughout the entire study as necessary.

Particle behavior is explained starting with simple steady-state, single-direction movement. The techniques used here are then expanded to include more complex motion and to relate how various forces influence particle behavior. Basic collection mechanisms are discussed as necessary to provide examples that show how fine particles are collected by various devices.

Particle size evaluations must be made before type of removal mechanism can be considered and to determine effectiveness of particle capture. In addition, particle-sizing devices are useful as further examples of collection mechanisms. Therefore, a brief but important discussion of size measurement is provided.

Construction of simple devices to study particles and the gaseous media can be useful in understanding the physical principles involved. The experiments provided are intended for this purpose and to suggest how simple but valuable devices can be made in emergency situations. Scientists need to be resourceful. More than once, loss of a device while on a test in a remote location and/or on a low-budget project has made it necessary for me to construct specific pieces of "on-the-spot" test equipment.

The last chapter includes some sample problems that can be used to test reader comprehension of much of the material presented. Some problems are workable after reading the first chapter while others should not be attempted until the relevant section is covered.

Howard E. Hesketh, P.E.
Carbondale, Illinois

ACKNOWLEDGMENTS

One of the rewards of writing a book is the opportunity to formally recognize the help received. I appreciate Ruth Schaffner Korte for her diligent preparation of the manuscript; Terry A. Sweitzer, my 1977 TEE-416 students and Joyce for helping proofread; and Billy S. Leasor for drawing help. It was possible to complete this work because of the time and help granted by Southern Illinois University at Carbondale.

DEDICATION

To my parents, Howard B. and Musetta E. Hesketh

CONTENTS

1. **Introduction**.   1

   1.1 Fine Particles   1
   1.2 Size and Size Distribution   2
   1.3 Diameter and Shape Relationships   7
   1.4 Properties 10
   1.5 Origination and Terminal States 13
   1.6 Dimensionless Numbers 16
   1.7 Size Regimes 20

2. **Unidirectional Motion of Particles** 23

   2.1 Defining Equations 24
   2.2 Gravitational Force 26
   2.3 Other Forces 30
   2.4 Particle Diffusion 30
   2.5 Aerodynamic Diameter. 34
   2.6 Shape Effects 37
   2.7 Particulate Clouds 39
   2.8 Effects of Adjacent Surfaces 42

3. **Nonsteady-State Motion of Particles** 45

   3.1 Characteristic Time 45
   3.2 Nonsteady-State Behavior 46
   3.3 Gravitational Settling of Particles 52
   3.4 Diffusive Deposition 55
   3.5 Curvilinear Motion. 59
   3.6 Scale-Up of Systems. 64

4. Effects of Other Forces **67**

    4.1 Inertial Force 67
    4.2 Electrostatic Force 74
    4.3 Magnetic Force 89
    4.4 Phoretic Forces 90
    4.5 Diffusion Force 96
    4.6 Acoustic Force 97
    4.7 Adsorption Force 100
    4.8 Other Forces 100

5. Particle Collection **107**

    5.1 Impaction and Interception 107
    5.2 Filtration 107
    5.3 Electrostatic Precipitation 111
    5.4 Collection Devices 115
    5.5 Mist Elimination 136
    5.6 Pressure and Temperature 138
    5.7 Interaction of Particles 140
    5.8 Cut Diameter 146

6. Size Measurement **151**

    6.1 General 151
    6.2 Cascade Impactors 156
    6.3 Diffusion Batteries 166
    6.4 Condensation Nuclei Counters 169
    6.5 Optical Counters 169
    6.6 Electrical Analyzers 171
    6.7 Aerosol Generation 172

7. Experiments and Problems **181**

    7.1 Experiments 181
    7.2 Problems 189

Appendix **199**

    Table of Nomenclature 199
    English and Metric Equivalents 205

Index **207**

CHAPTER 1

INTRODUCTION

1.1 FINE PARTICLES

The behavior and characterization of very small particulate matter in gaseous media has become increasingly important over the past decade because of the effects of air pollution and the desire to control these emissions. The expression "fine particles" has been established because of the confusion in the use of terms such as aerosols, smokes, fumes, dusts, mists and clouds. The currently accepted meaning of fine particles in the U.S. and the one used in this book is "particulate matter with an aerodynamic diameter of 3 micrometers (μm) or smaller." Theoretically, particles as small as several molecules can exist; however, the practical lower limit for control is about 0.01 μm. These discussions deal specifically with particles in gaseous media, and the particulate matter may be solid and/or liquid. Methods for predicting behavior of particles in liquid fluids are similar to those used for gaseous fluids, and use of proper fluid physical constants is usually all that is required to calculate behavior of particles in liquid media. However, appropriate factors may have to be considered to account for the particle Reynolds number variations (because particles move more easily through gaseous media) and for the difference in particle-fluid interactions; *e.g.,* liquid particles may be soluble and solids may dissolve in the liquid.

It is important to understand the behavior of fine particles to control it, as the material has a significant effect on our lives. As air pollutants, fine particles comprise by number the greatest amount of emissions. Yet this material is the most difficult to control and can be the most detrimental to our health when inhaled. Particles larger than 3 μm in diameter can be removed with virtually 100% efficiency by many standard medium-high energy control devices. When particles are breathed in, the natural body defense mechanisms prevent most larger particles from reaching the lower respiratory tract. It is the particles in the 0.2-0.5 μm size range that are most detrimental to the lungs although particles as small as 0.01 μm are deposited. These

1

control- and health-related factors are now being recognized and in certain geographical areas industrial growth is being prohibited, even though existing air pollution health standards are not expected to be exceeded and particulate control equipment is planned.

Particles are often assumed to be spherical and rigid because it is easiest to predict the behavior of this type of material. This is a convenient first approximation but must be corrected when such is not the case. Unless otherwise noted, most sections of this book deal with the simpler cases of spherical particles. The sections that discuss how other shapes can be accounted for must be applied to the basic spherical procedures whenever necessary. Also, note that behavior of particles larger than 3 μm is accounted for by many of the procedures given.

1.2 SIZE AND SIZE DISTRIBUTION

Particulate matter can be considered either monodisperse or polydisperse. Dispersion implies amount of particle size spread. Monodisperse means that all particles in the sample are essentially the same size while polydisperse means there is a distribution of size. All particulate matter in gaseous media is polydisperse to some degree; however, it is sometimes convenient to assume that the particles are of some average or mean diameter. This is only a simplifying assumption; for accurate predictions, detailed corrections may be made to account for each particle size.

Polydisperse particles usually have a natural or log-normal distribution. Cumulative distribution plots of most of these materials yields a straight line when log of the diameter is plotted versus cumulative mass, area or number on a probability scale (Figure 1.1). The details of this procedure have been described.[1] An alternate procedure using reversed indices has been shown by Fuchs.[2]

The data of Figure 1.1 are emissions from a fuel oil combustion boiler. By number, 6.1% are greater than 5 μm, 16.7% are from 2-5 μm, 28.8% are from 1-2 μm and 48.4% are 1 μm or less. These values, plotted as cumulative % undersize, give the solid line in Figure 1.1. Note that size distribution by mass, area and number are related by having essentially the same slope, and therefore, the same geometric standard deviation (σ_g). This relationship depends on actual particulate shapes and densities. These plots indicate accuracy of data and condition of the sampled particulate matter. The dotted curves in Figure 1.1 are calculated assuming consistent spherical shape and density. A *suggested* equation to relate mass mean, $\bar{d}_M$, and number median, $\bar{d}_N$, diameters is $ln\ \bar{d}_M/\bar{d}_N = C_1\ ln^2\ \sigma_g$ where the constant C_1 is about 2.66 but needs to be verified by further work. Actual particulate sampling techniques are discussed in Chapter VI.

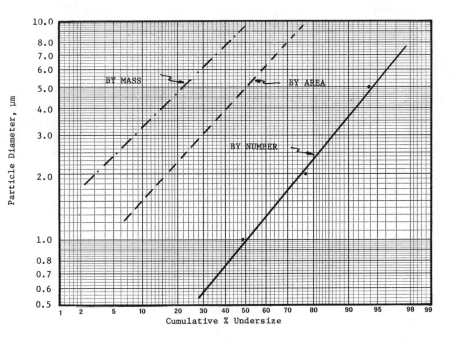

Figure 1.1   Log-probability cumulative distribution of emissions from a fuel oil boiler.

The σ_g can be obtained mathematically, or more simply, it can be obtained directly from the plot

$$\sigma_g = \frac{d_{84.13}}{\bar{d}} = \frac{\bar{d}}{d_{15.87}} \tag{1.1}$$

The values of $d_{84.13}$, $\bar{d}$, and $d_{15.87}$ are the diameters that occur on the plot at the cumulative probability distribution of 84.13, 50 and 15.87% respectively. The mean diameter $(\bar{d})$ must be specified as to whether it is the mass, area or number mean. The mean diameter by number of the sample in Figure 1.1 is 0.98 μm and σ_g is 2.8. The mean diameter by area is 5.0 μm and the mass mean diameter (MMD) is 9.5 μm.

The mass mean could also be found by dividing the mass of a specific unit volume by the number of particles in that mass. From that, MMD for spherical particles is

$$MMD = \left(\frac{\text{mass mean}}{(1/6)\pi\rho_p}\right)^{1/3} \tag{1.2}$$

where ρ_p is density of the particle.   For other than spherical particles an "equivalent" diameter would be obtained.

Improvements in sampling and measurement techniques reveal that many particulates have bimodal distribution.   The particulate sample may be a single material with two or more size distributions and/or a group of several materials with several size distributions. A single substance, for example, *part* of which has been fractured (mechanically, thermally, etc.) can result in a sample with two size distributions. Alternatively, two or more materials of varying size distributions can coexist.   Figure 1.2 shows bimodal fly ash containing unburned coal dust and ash.   Analysis showed the larger (darker) particulates to be coal.   The ash was smaller and nearly white.   Figure 1.3 shows a similar situation for ambient particulates.   In this latter case it is

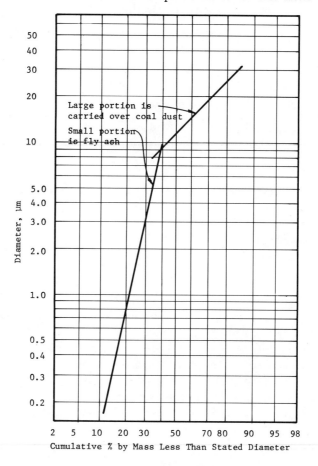

Figure 1.2   Aerodynamic size measurements of fly ash from a stoker boiler.

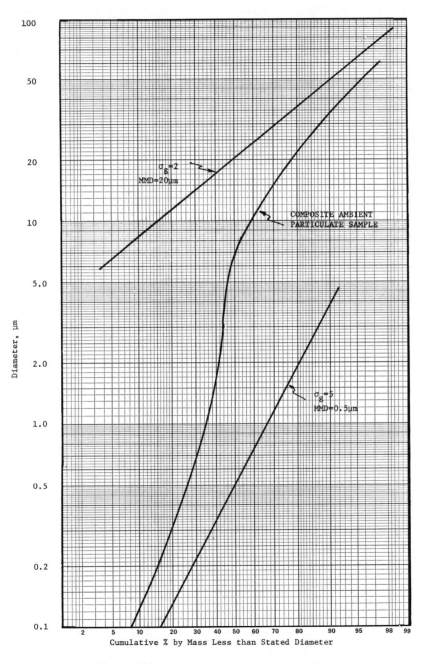

Figure 1.3 Ambient particulate matter sample.

believed that a smaller condensation aerosol (σ_g = 5) formed the lower portion of the composite curve and a larger dispersion aerosol (σ_g = 2) formed the upper.

Degree of polydispersity can be estimated using the relative standard deviation (α). For values of $\alpha < 0.1$, the particulate sample can be considered to be monodisperse. Relative standard deviation is

$$\alpha = \frac{\sigma_g}{\bar{d}} \tag{1.3}$$

Figure 1.4 shows schematically how values of α change using typical frequency distribution curves of particles with a similar median (middle of sample by number) diameter size. Notice that log-normally distributed material produces a frequency distribution curve that is skewed to the right when plotted on cartesian coordinates.

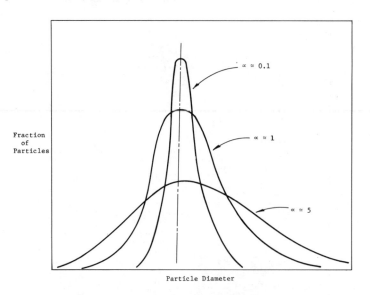

Figure 1.4 Typical frequency distribution curves.

Glass furnace effluent consists of extremely small particulates with a $\bar{d}$ of about 0.5 μ and σ_g = 1.6. Is it monodisperse or polydisperse?

$$\alpha = \frac{1.6}{0.5} = 3.2$$

showing it is very polydisperse.

1.3 DIAMETER AND SHAPE RELATIONSHIPS

It is convenient to express the behavior of an irregularly shaped particulate specimen as if it were a spherical particle. Use of such common denominators makes it easier to predict, compare and correlate various material. The term aerodynamic diameter was used in Section 1.1 in relation to the definition of fine particles and equivalent diameter was used in Section 1.2 when discussing mean diameters. Definitions for these and other useful expressions follow.

1.3.1 Aerodynamic Diameter (d_a)

In situations where gravitational or inertial forces act on a particle proportional to its mass, aerodynamic diameter is a "pseudo" particle size which specifies its motion. Stober[3] defines aerodynamic diameter as the diameter of a sphere of unit density ($\rho_p = 1$ g/cm^3) that attains the same terminal settling velocity (v_s) at low Reynolds number in still air as the actual particle under consideration. This diameter is of physical interest because often it can be obtained by a direct dynamic measurement of the particle in motion where other measures of actual particle size and density are not obtainable. The mathematical procedure necessary to obtain true diameter from aerodynamic diameter is given in Section 2.5.

1.3.2 Equivalent Diameter (d_e)

This is the diameter of a sphere having the same volume as the particle under consideration. It can be obtained by

$$d_e = \left(\frac{6V}{\pi} \right)^{1/3} \tag{1.4}$$

where V is the volume of the particle. If the volume of the entire sample divided by the number of particles in the sample is used for the volume term in Equation (1.4), then the sample average d_e is obtained.

1.3.3 Sedimentation Diameter (d_s)

This is the diameter of a sphere having the same terminal settling velocity and density as the particle under consideration. Reduced sedimentation diameter is the same except that the sphere density is unity, making it the same as aerodynamic diameter. For spherical particles $d_s = d_e$; for others, these two diameters can be related by the dynamic shape factor (χ), as noted in Section 1.3.6.

1.3.4 Projected Diameter (d_o)

This is the diameter of a circle whose area is equal to the projected maximum cross-sectional area of the particle as seen from a specified angle perpendicular to the direction of flow. It can be obtained from microscopic examination of the particle. Optical microscopes may have to be focused at each particle to obtain this maximum area, but electron microscopes with greater depth of field or holographic microscopy can give this directly for an entire specimen. For example, when taking a picture of fine particles using an optical microscope with a numerical aperture value of 1.3, the photographic depth of field in white light is only about ¼ μm.

1.3.5 Cut Diameter (d_c)

Cut diameter is a valuable single parameter used to describe the efficiency of particle collection devices. It is the particle diameter for which the efficiency (E) and the penetration (P) is 50%; *i.e.*, half these size particles are captured and half penetrate or escape from the collector. Note that in symbols

$$P = 1 - E \tag{1.5}$$

where both P and E are expressed as fractions.

1.3.6 Dynamic Shape Factor (χ)

The equivalent and sedimentation diameters are related[2] by the dimensionless dynamic shape factor:

$$\chi = \left(\frac{d_e}{d_s} \right)^2 \tag{1.6}$$

We have already noted that d_e and d_s are equal for spherical particles, so χ for spheres is 1.0. For nonspherical particles, measurements reveal that equivalent diameters are nearly always larger than sedimentation diameters, so normally χ is > 1.0.

Numerical values of χ can be predicted for various shapes using Pettyjohn's procedure:[4]

$$\chi = \left(0.843 \log \frac{\Psi}{0.065} \right)^{-1} \tag{1.7}$$

Values of the dynamic shape factor using the noted values of sphericity factor (Ψ) are given in Table 1.1 for various shaped particles at low particle Reynolds number (< 0.1).

Table 1.1 Dynamic Shape (χ) and Sphericity (Ψ) Factors

Particle Shape	Ψ	χ
Sphere	1.000	1.00
Cubical octahedron (8 equal sides— 4 triangles each on top and bottom)	0.906	1.04
Octahedron or rodlet	0.846	1.06
Cube or regular rectangle	0.806	1.08
Tetrahedron or splinter	0.670	1.17

Large particles or aggregates of small particles will fall at a faster rate and can have Reynolds numbers > 0.1. Under these conditions, the particles usually become oriented so as to provide the maximum resistance to fall; that is, their major axis is perpendicular to the fall direction. Data on glass sphere aggregates suggest how χ and $(d_e/d_s)^2$ increase with aggregate size (Figure 1.5).

A particle of given shape will have a dynamic shape factor as predicted by Table 1.1, for example, if the particle is small. As particle sizes increase, Reynolds numbers increase in free-fall situations resulting in an increase in χ. Therefore, a larger particle of the same given shape could be expected to have a larger value of χ.

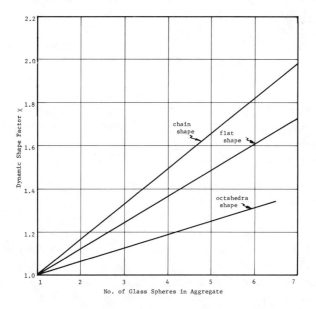

Figure 1.5 Effect of agglomeration on dynamic shape factor.[5]

1.4 PROPERTIES

This section discusses some specific properties of fine particulate matter as individual particles and as classified by various group names. Detailed presentation of certain properties, such as electrical charges, will be covered in later sections.

1.4.1 Density

Densities of fine particles may be considerably different than those of large pieces of the same material. True densities of elements and various pure compounds are available in standard handbooks. The measured apparent densities of fine particles in gaseous media are usually lower than the reported values, often from 2 to 10 times lower. There are several explanations for this.

The previous section described how the dynamic shape factor varied upwards from 1.0 as irregularity of shape and/or agglomeration increased. This, in effect, is indicating that settling rate, which is a function of particle density, is decreasing. It is suspected that density values measured by settling or aerodynamic techniques that are about one-half the true value occur when the particle is single and densities about 1/10 the expected value are aggregates of particles.

Smaller particles have a greatly increased surface area per unit volume. This increases the chance of reactions, especially oxidation, because of the normal presence of oxygen in the gaseous media. If the original particulate were a metal, most oxides and compounds would have a lower density than the original material. Gases can be adsorbed on the large surface area per unit mass of very small particles, which can result in apparent particle density-size variations. In moist conditions, hygroscopic particles often become more compact as they solubilize. This can increase the apparent density.

Solid particles may not be solid throughout. Fly ash particles, called cenospheres, have hollow centers, and this decreases apparent density.

1.4.2 Charge

As aerosols, fine particulates can have no charge, a positive charge or a negative charge. Gaseous ions are often confused with very small charged particles because both have about the same diffusion velocity. When ions combine, the products have neutralized charges, by contrast with charged larger particles resulting when particles agglomerate (solids) or coalesce (liquids). Most particles larger in diameter than 0.1 μm contain one elementary charge (1 electron) as a result of natural diffusion charging. Few smaller particles have charges. As particle size increases, the chances of having more than one elementary charge increases. The amount of charge affects particle electrical mobility and behavior in electric and magnetic fields.

1.4.3 Optical Effects

Particles in gaseous mediums may be transparent, semitransparent or opaque. Each of these properties can affect particle movement differently. Light striking a transparent object may pass through the object and warm the far side. This causes gas molecules near the warmed surface to heat up and push the particle toward the light. An opaque object by the same procedure may be warmed on the near side and move away from the light. Semitransparent particles may do either.

Particles scatter light proportional to their size. Those larger than about 1 μm scatter light proportional to their diameter squared; smaller ones scatter it proportional to diameter to the sixth power. For this reason, particles smaller than 1 μm cannot usually be seen by optical microscopes in white light. The theoretical limit of a microscope to resolve two discrete points separated by a distance x is

$$x = \frac{1.22\lambda}{2NA} \qquad (1.8)$$

where λ is light wavelength and NA is numerical aperture of objective. Using λ of white light as 0.45 μm (4500 Angstroms) and the best aperture for visible light (NA $\cong$ 1.3) shows that two particles 0.1 μm apart could be seen as being separate. Under the best conditions, this would require at least 300 magnification; under less optimum conditions as much as 1500 may be needed.

Scattering of sunlight by atmospheric aerosols results in elimination of the high-frequency colors (blue end of the spectrum) and only the low frequency reds are able to pass. Brilliant red sunrises and sunsets may be observed because of this on days when the air is heavily polluted.

1.4.4 Surface Variations

Noticeable variations from the normal are observed as liquid particles decrease in size. Defay[6] and others have shown that surface tension decreases and apparently causes vapor pressure to increase and heat of vaporization to decrease. This is shown for water droplets in Figure 1.6.

Large amount of surface area per unit mass, i.e., specific surface, is a characteristic feature of fine particles, and this causes problems in working environments and pollution control systems. Measured surface areas are usually found to be greater than calculated areas because geometrical

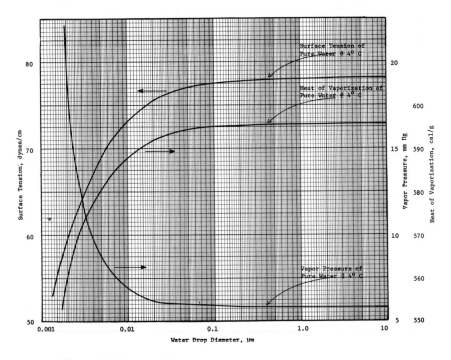

Figure 1.6 Surface effects as related to particle size of water droplets. From Defay.[6]

evaluations do not include factors to account for surface roughness or interstitial surface variations. Foust *et al.*[7] present data to show how specific surface is a function of particle size. This is plotted for quartz and pyrite in Figure 1.7.

1.4.5 Explosiveness

Material that is normally safe can become extremely explosive as particle size decreases. Jacobson *et al.*[8] showed that 100 μm aluminum particles were weakly explosive while 10 μm particles were strongly explosive. The minimum igniting energy increases in a direct logarithmic relationship from about 30 millijoules (mj) for 10 μm particles to about 200 mj for 100 μm particles. Examples of the explosive nature of several types of small particles are given in Table 1.2. These are average values from Hartmann[9] and would fluctuate with size variations.

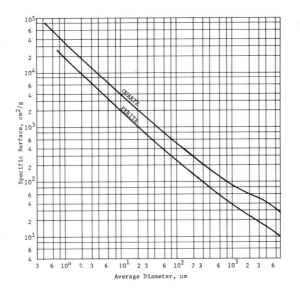

Figure 1.7 Measured specific surface as a function of particle size.[7]

1.5 ORIGINATION AND TERMINAL STATES

Atmospheric particulates can be considered primary or secondary particulates. Primary particles are those emitted into the atmosphere in essentially final form while the secondary particles are formed in the atmosphere by reaction of primary particulates and/or gases or by condensation. Secondary particles are usually much finer in size than the bulk of primary material and hence, tend to remain suspended in the atmosphere for long periods of time. However, very fine primary particulates also remain suspended. Other terms can also be used such as dispersion aerosol (material emitted directly as a result of mechanical action) and condensation aerosol (formed when supersaturated vapors condense). These terms are similar to primary and secondary particulates but do not include all the possible methods of formation.

Clouds of particulate matter in the atmosphere have been given names depending mainly on their mechanical or natural origin. Explanations and definitions of some traditional terms are given in the following paragraphs, and the definition of fine particles is given on page 1 of this chapter. Most of these definitions are relatively general. Note that this includes those portions of particulates defined as primary or secondary particulates that have an aerodynamic diameter of ≤ 3 μm. The same holds true for particulates

Table 1.2 Explosive Character of Dusts[9]

Dust	Ignition Temperature (°C)	Minimum Spark Energy for Ignition (mj)	Minimum Explosive Concentration (g/m³)	Maximum Explosion Press (atm)	Average Rate of Pressure Rise (atm/sec)	Oxygen to Prevent Ignition (<%)
Aluminum, atomized	640	15	40	6.1	240	7
Magnesium, atomized	600	120	10	5.4	140	3
Cellulose acetates	320	10	25	7.5	190	11
Polystyrene	490	15	15	6.1	160	14
Rubber, synthetic	320	30	30	6.5	75	15
Alfalfa	460	320	100	~ 4.4	34	–
Clover seed	470	80	60	4.1	27	15
Coffee	410	160	85	3.4	10	13
Cotton seed	470	80	55	6.1	54	15
Wheat	470	50	70	7.1	100	all
Cellulose	480	80	55	6.8	75	13
Soap	430	60	45	5.8	41	all
Wood flour	430	20	40	7.5	110	17

classified by the following categories. It has already been pointed out that atmospheric particulate matter usually consists of two or more basic types of material and even direct emission particulates are bimodal, often as a result of containing two or more types of matter. Background particulate matter (natural, man-made or both), locally generated emissions (natural, man-made or both) and secondary particulate products are always present in the atmosphere.

Mists — These are suspensions of liquid droplets in the atmosphere formed by condensation of vapors or resulting as direct dispersion aerosols formed by atomization, for example. Most of these particles are larger than fine particles as they have diameters of 5 μm and larger. Fog is considered to be a high concentration of mist.

Dust — This consists of dispersion aerosols of solid particulate matter up to several hundred μm in diameter. The origin of this material could be from naturally occurring wind-generated actions or from industrial processes.

Smokes — Smokes include both solid and liquid particulates, the latter usually being condensation aerosols formed by condensation from supersaturated vapors.

Fumes — Fumes are a type of smoke considered to be the condensation of metal or metal oxide vapors.

Hazes — This is a "catch-all" category including both mists and dusts. Condensed water vapor is usually a significant contributor.

Smog — Smog is a contraction of the words smoke and fog, and is usually found over metropolitan communities during times of atmospheric inversion.

Aerosols — Aerosols or aerocolloidal systems have been given various definitions usually consisting of particles less than 50 μm in diameter; however, the use of the term aerosol sprays has caused much confusion. Particles developed by aerosol spray cans are much larger in size, so usually cannot be considered to be aerosols as we define them. Hidy[10] defines aerocolloidal system as those particles that have a Reynolds number less than 1 and a surface-to-volume ratio greater than 10^3 cm^{-1}. With this definition, systems ranging from extremely high concentrations such as fluidized beds and systems with extremely dilute concentrations such as interplanetary dust can all be considered aerocolloidal systems.

C + E News Aug Nov Sept 80 says that Mt St Helens smoke is particles heavily laden with H2SO4 during eruption

1.6 DIMENSIONLESS NUMBERS

Various groupings of physical parameters arranged so as to be dimensionless are useful to identify specific properties of a given system. Some of these numbers can be considered ratios of forces or effects in the system; these numbers are necessary in this study, and already the Reynolds number has been mentioned several times. Some numbers used in the study of particulates in gaseous mediums will be reviewed and explained in the following summary. Use and further explanatory information will be given later, in the appropriate sections. The dynamic shape and sphericity factors are also dimensionless and are discussed in Section 1.3.6.

1.6.1 Flow Reynolds Number (Re_f)

This is an extremely useful relationship related to fluid flow. Reynolds used the number to obtain critical values of velocity where flow changes from laminar to turbulent. The equation is

$$Re_f = \frac{D \, v_g \rho_g}{\mu_g} \qquad (1.9)$$

where D = diameter (or equivalent) of containing device
v_g = velocity of gas or fluid
ρ_g = density of gas or fluid
μ_g = viscosity of gas or fluid

Roughly speaking, fluid flow is turbulent when the Reynolds number is greater than 4000 and flow is viscous below 2100, although there are frequent exceptions to this depending on existing conditions. Flow friction is a function of Reynolds number. This number can be considered as the ratio of *inertial* to *viscous* forces when arranged as

$$Re_f = \frac{\rho_g v_g^2 / D}{\mu_g v_g / D^2} \qquad (1.10)$$

1.6.2 Particle Reynolds Number (Re_p)

A particle flow expression is needed when the particle moves in an unconfined fluid medium or in a system where the walls of the device have essentially no effect on the particle. The particle Reynolds number used for this is

$$Re_p = \frac{d(v_p - v_g)\rho_g}{\mu_g} \qquad (1.11)$$

where d = diameter of particle
v_p = velocity of particle
v_g = velocity of gas or fluid

In contrast to the large numerical values noted for flow Reynolds number, *[very interesting statement]* particle Reynolds number for fine particles in gaseous mediums is typically less than 0.1.

1.6.3 Knudsen Number (Kn)

There are times when a fluid no longer behaves as a continuum. The Knudsen number[11] can be used to indicate how far a gaseous system, including the particles, deviates from the normal. This number is defined for any positive number and zero. As Kn becomes smaller, the gas acts more normal and behaves as a continuum. The number is

$$Kn = \frac{2\lambda_g}{d} \tag{1.12}$$

where λ_g is the mean free path of the gas molecules. According to this definition, Kn can vary for a given size particle depending on what gas is present and whether a gas is hot or cold.

The kinetic-molecular theory predicts that gas mean free path can be estimated using

$$\lambda_g = \frac{\mu_g}{0.499\, \rho_g} \sqrt{\frac{\pi M}{8RT}} \tag{1.13}$$

where M = molecular weight
R = universal gas constant
T = absolute temperature, $^{\circ}K$

In metric units, the value of R needed to clear dimensions would be 83.12 X 10^6 g cm^2/(sec^2 gmole $^{\circ}$K). For air at standard conditions (SC) of 1 atmosphere and 20°C the mean free path predicted by Equation (1.13) is about 6.60 X 10^{-6} cm.

Adapting the kinetic theory equation to the Glasstone[12] viscosity relationship gives a simplified expression that can be used for air near standard conditions:

$$\lambda_{air} = (2.26 \text{ X } 10^{-5})(T/P) \tag{1.14}$$

where P is the pressure in millibars (mb). At SC (1013 mb), the mean free path of air is about 6.53 X 10^{-6} cm (6.53 X 10^{-2} μm). We will use this value, although Wilson[13] reports a value of 8 X 10^{-6} cm. Fortunately in air pollution control the gas is often air or a mixture very similar to air.

1.6.4 Mach Number (Ma)

Mach number is simply the ratio of gas velocity (v_g) to the acoustic velocity (v_a), which is the velocity of sound in the fluid at the local conditions.

$$Ma = \frac{v_g}{v_a} \tag{1.15}$$

The speed of sound in an ideal gas is

$$v_a = \sqrt{\frac{\gamma RT}{M}} \tag{1.16}$$

where γ is the ratio of specific heats (C_p/C_v), which equals 1.40 for ideal gases.

Substituting Equation (1.13) for the value of λ_g in Equation (1.12), and assuming that the particle velocity is negligible compared to gas velocity and neglecting signs, we can show that

$$Ma \cong \frac{1}{4}(Re_p)(Kn) \tag{1.17}$$

For this reason both the Reynolds and Knudsen numbers can be considered independent variables.

As long as Ma is $\ll 1$, the gas experiences incompressible flow. As Ma increases in value, considerable changes in density, temperature and pressure occur, and compressible flow must be considered. This work considers only incompressible flow, unless otherwise stated.

1.6.5 Schmidt Number (Sc)

This number considers the diffusivity of particles and for isothermal mass transfer is analogous to the Prandtl number in heat transfer. The Schmidt number correlates the properties of viscosity, density and diffusivity whereas the Prandtl number correlates specific heat, viscosity and thermal conductivity. Both are relatively independent of pressure and temperature when conditions are near SC. The Schmidt number is

$$Sc = \frac{\mu_g}{\rho_g D_{PM}} \tag{1.18}$$

where D_{PM} is the diffusivity of the particle through the gas (see Section 2.4). Brownian diffusivity becomes less significant compared with convective mass transfer of particles, as Sc increases in value.

1.6.6 Stokes Number (St)

Stokes number is important in describing the particle collection potential of a specific system. It is the ratio of stopping distance of a particle to the distance a particle must travel to be captured.

$$St = \frac{2x_s}{D} \tag{1.19}$$

where x_s is the particle stopping distance.

1.6.7 Froude Number (Fr)

This is similar to the Stokes number but is derived considering gravity as the particle-gas separating force.

$$Fr = \frac{v_g^2}{D\,g}$$ (1.20)

where g is the acceleration of gravity.

1.6.8 Impaction Parameter (K_I)

This factor is also similar to the Stokes number and in some cases is identical. There are several ways this factor can be expressed, so to minimize errors it is best to check the author's definition each time it occurs. Several forms of K_I are in Section 5.1.

1.6.9 Cunningham Correction Factor (C)

Particles that fall in the "Slip Flow Regime" can often be treated in the same manner as larger particles, if Cunningham correction is applied. For this reason, it is extremely useful and important. Cunningham[14] first proposed this correction in the form of

$$C = 1 + A\,\frac{2\lambda_g}{d}$$ (1.21)

where A is a "constant." It is shown[12] that at standard pressure of 1 atmosphere, the value of C for particles in air can be found from

$$C = 1 + \frac{(2 \times 10^{-4})\,T}{d_p}\,[2.79 + 0.894 \exp(-\frac{2.47 \times 10^3\,d_p}{T})]$$ (1.22)

where d_p is particle diameter in μm.
A simplified version[15] can be used to obtain good estimations:

$$C = 1 + \frac{(6.21 \times 10^{-4})\,T}{d_p}$$ (1.23)

For gases other than air and for particles with Kn greater than about 1, Davies[16] suggests the value of A in Equation (1.21) be determined by

$$A = 1.257 + 0.400 \exp(-1.10/Kn)$$ (1.24)

1.6.10 Peclet Number (Pe)

The combined effects of diffusion and fluid motion (convection) on particle transport can be expressed as a function of Peclet Number (Pe), which is

$$Pe = \frac{v \, D_c}{D_{PM}} \qquad (1.25)$$

where D_c is collector diameter and v is particle average drift velocity.

1.6.11 Other Dimensionless Numbers

When used, other dimensionless numbers are less general in nature, and specific defining equations are given as needed in the appropriate sections.

1.7 SIZE REGIMES

One set of mathematical relationships cannot predict particle dynamics for all sizes of particles and under all conditions. Indeed, there are four separate size regimes that can be considered for fine particles in gaseous media near standard conditions. Ranging from small to large particles in order, these are: (a) Free Molecule; (b) Transition; (c) Slip Flow; and (d) Continuum. At this time there has been no generally acceptable procedure developed to analytically describe particle transport processes in the Transition regime. The reader will find there are times when the Slip Flow Regime is also named the Cunningham Regime and the Continuum Regime is known as Stokes Regime. The names Stokes and Cunningham are often applied to these regimes in recognition of the significant contributions these men have made in defining particle behavior in the specified regimes. Actually, *variations* of Stokes' procedure are better for larger and smaller particles in the Continuum, and Cunningham's procedure may be used for certain size particles in all regimes. Keeping in mind the provisions noted, we will use these terms interchangeably for the respective regimes.

In air pollution control the relative motion between particles and gases is usually quite small, and gas velocities are usually much below acoustic velocities (*i.e.*, Ma $\ll$ 1). Under these conditions we consider the gas incompressible. Other common assumptions are constant gas density and composition. Note that if significant variations make these stated conditions variables, then additional parameters should be considered to more correctly account for particle behavior. Conditions that can produce such effects are, for example: high gas velocity, large temperature and/or concentration gradients resulting from wet scrubbing or dilution operations, high pressure and passage through mechanical devices such as nozzles.

Under the simplifying assumptions, fine particles in gaseous media will usually have a Ma $\ll$ 1 and a $Re_p < 1$. With these constrictions, the four major size regimes can be denoted by the Knudsen number as given in Equation (1.12). Using the mean free path of air near standard conditions of 6.53 X 10^{-6} cm as calculated by Equation (1.14), an approximate relation between Knudsen number and particle diameter for the four size regimes is given in Table 1.3. Values of Kn are > 10 in the Free Molecule Regime, 0.3-10 in Transition Regime, ≤ 0.3 in the Slip Flow Regime and very small in the Continuum Regime.

Table 1.3 Approximate Values of Knudsen Number and Particle Diameter for Various Size Particles in Air at SC

	Size Regimes			
	Free Molecule	Transition	Slip Flow (Cunningham)	Continuum (Stokes)
Kn	>10	10-0.3	<0.3	<0.1
d, μm	<0.01	0.01-0.4	>0.4	>1.3

Figure 1.8 shows how Fuchs[2] predicts the accuracy by which particle dynamics can be predicted by available formulas and data for various-sized spherical particles of unit density in air, under normal conditions. For comparison, the approximate regimes obtained from Kn and λ_{air} are located by particle size on Figure 1.8. Note the gap for particles falling in the Transition Regime. Each regime will be considered separately in the next chapter except for the Transition Regime, which must be considered part of the Free Molecular or Slip Flow Regimes.

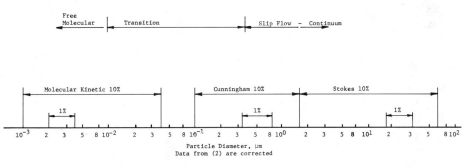

Figure 1.8 Schematic of approximate regime sizes for particles in normal air.

REFERENCES

1. Hesketh, H. E. *Understanding and Controlling Air Pollution* (Ann Arbor, Michigan: Ann Arbor Science Publishers, Inc., 1974).
2. Fuchs, N. A. *Mechanics of Aerosols* (London: Pergamon Press, 1964).

3. Stober, W. "A Note on the Aerodynamic Diameter and Mobility of Nonspherical Aerosol Particles," *J. Aerosol Sci.* 2:453-456 (1971).
4. Pettyjohn, E. S. and E. B. Christianson. "Effects of Particle Shape on Free-Settling Rates of Isometric Particles," *Chem. Eng. Prog.* 44(2): 157-172 (1948).
5. Kunkel, W. J. *J. Appl. Phys.* 19:1056 (1948).
6. Defay, I. *et al. Surface Tension and Adsorption* (New York: John Wiley & Sons, Inc., 1966).
7. Foust, A. S., L. A. Wengel, C. W. Clump, L. Maus and L. B. Anderson. *Principles of Unit Operations* (London: John Wiley and Sons, Inc., 1960).
8. Jacobson, M., A. R. Cooper and J. Nagy. "Explosibility of Metal Powders," U.S. Dept. of Interior, Bureau of Mines Report of Investigation, ROI 6597 (1965).
9. Hartmann, I. "Dust Explosions," In *Marks' Mechanical Engineering Handbook,* T. Baumeister, Ed. (New York: McGraw-Hill Book Co., 1965).
10. Hidy, G. M. and J. R. Brock. *The Dynamics of Aerocolloidal Systems* (London: Pergamon Press, 1970).
11. Knudsen, M. *The Kinetic Theory of Gases* (London: Methuen Publishers, 1934).
12. Glasstone, S. *Textbook of Physical Chemistry* (New York: Van Nostrand and Co., 1946).
13. Wilson, L. G. and P. Cavanagh. "The Capture of Sub-Micron Aerosol Particles by Single Fibres," *Atmos. Environ.* 5(3):123-136 (1971).
14. Cunningham, E. *Proc. Royal Soc.* 83A:357 (1910).
15. Calvert, S. *et al. Scrubber Handbook* (APT, Inc., 1972).
16. Davies, C. N. *Proc. Phys. Soc.* (London) 15(61) (1943).

CHAPTER 2

UNIDIRECTIONAL MOTION OF PARTICLES

Particle dynamics are presented in this chapter for each size regime, in order, from the largest to the smallest particles: Continuum Regime, Slip Flow Regime and Free Molecule Regime. The Transition Regime, which lies between the Slip Flow and Free Molecule Regimes, will not be covered as a separate case, as it is best covered by extension of the bounding regimes procedures. The effects of various forces acting on particles will be considered starting with simplified cases then combined effects. The three basic types of forces that can act in any and all combinations are (a) external, (b) resistance of medium and (c) interaction. External forces in air pollution studies always include gravitational force and may include other forces such as electrical, magnetic, thermal or nuclear. In studies of particles in fluid media, the medium resistance on a particle is always present whether the particle is in motion or at rest. This force is not considered an external force because the medium is always present and, being continuous, it usually is not isolated. Interaction of particles in relation to particle motion is minor in systems of dilute particle concentrations such as in the atmosphere. Particle interaction effects become more significant in source emission streams and in collection devices where number concentrations are high.

Regime limits are noted in Chapter 1 based on Knudsen number values at low Mach and particle Reynolds numbers. These cannot be considered as exact values because much of the work in this field is empirical correction of theoretical approximations.

The case of simple, *monodispersed spherical* particles in terminal *steady state* motion in a *gravitational* field is discussed first, followed by effects of other forces and influencing conditions. As noted in Chapter 1, it is assumed for this homogeneous gaseous fluid that motion of both gas and particulate matter is sufficiently slow so Mach number (Equation 1.15) is $\ll 1$ and drop Reynolds number (Equation 1.11) is properly related to the particle-gas

23

drag characteristics. Under these conditions both the particle and gas are considered noncompressible and the particle is also assumed to be rigid. Effects from the pressure of external surfaces (container walls or other particles) are considered later.

2.1 DEFINING EQUATIONS

Equations of change can be written to describe particle-gas system behavior. For these constant density and viscosity systems this includes equations of continuity, motion and energy. The equation of continuity describes the rate of change of density resulting from changes in the mass velocity vector, ρv. Vector notation in terms of the substantial derivative of density, $D\rho_g/Dt$, gives for this:

$$\frac{D\rho_g}{Dt} = -(\nabla \cdot \rho_g v_g) \qquad (2.1)$$

This form of the equation using the substantial operator D/Dt describes the rate of change of density as observed by one moving along with the fluid where t stands for time. The divergence term $(\nabla \cdot \rho_g v_g)$, sometimes written as "div ρv," means the net rate of mass flux per unit volume. The expansion of div ρv gives for the right side of Equation (2.1)

$$\frac{D\rho_g}{Dt} = -\rho_g (\nabla \cdot v_g) - (v_g \cdot \nabla)\rho_g \qquad (2.1a)$$

In our limiting conditions we specified that the fluid is assumed incompressible, so $(\nabla \cdot v_g) = 0$, and that steady state exists, so $(v_g \cdot \nabla) = 0$. Under these conditions the equation of continuity is zero.

The equation of motion is obtained from a momentum balance on the system of the form of Newton's Second Law of Motion: mass times acceleration equals sum of forces. Expressed in tensor notation this equation can be shown as

$$\rho_g \frac{Dv_g}{Dt} = -\nabla P - (\nabla \cdot \tau) + \rho_g E \qquad (2.2)$$

where P is pressure, E is external forces and τ is the stress tensor. The expression indicates the resultant force acting on a particle. The force terms represented by the right side of this equation symbolize per unit volume of element: pressure force (first term), viscous force (second term), and external forces (third term).

For a Newtonian fluid at constant density and viscosity, Equation (2.2) simplifies to the Navier-Stokes Equation:

$$\rho_g \frac{Dv_g}{Dt} = -\nabla P + \mu_g \nabla^2 v + \rho_g E \qquad (2.3)$$

Note that the left side of this equation can be expanded to show

$$\rho_g \frac{Dv_g}{Dt} = \rho_g \left(\frac{\partial v_g}{\partial t} + (v_g \cdot \nabla) \, v_g \right) \qquad (2.4)$$

Thus, the left side of Equation (2.3) is related to velocity squared. The second term on the right in Equation (2.3) contains the viscosity term and therefore is related to friction. As velocity approaches zero, the left side of Equation (2.3), the inertial side, decreases more rapidly than the friction term, which only contains a first-order velocity term.

For large systems where there is little change in the stress tensor or when viscous effects are relatively unimportant, the friction term in Equation (2.3) drops out, giving the Eüler Equation:

$$\rho_g \frac{Dv_g}{Dt} = -\nabla P + \rho_g E \qquad (2.5)$$

Equations of energy consist of making an energy balance on the system to obtain a partial differential expression that describes energy transport in a homogeneous fluid or solid. This includes energy in the form of mechanical, chemical, internal, kinetic, thermal and viscous dissipation. This is usually not significant for the study of fine particle behavior in gases, so this will be considered negligible.

Very slow flow of a fluid around a rigid spherical particle exerts both a normal and tangential force on the surface of the sphere. The force per unit area (pressure) of the fluid exerted on the sphere gives the buoyant force (F_B). This is equal to the displaced mass of fluid times gravitational acceleration (g), or in terms of particle diameter (d) is

$$F_B = 1/6 \, \pi d^3 \, \rho_g g \qquad (2.6)$$

Analytical evaluation of the momentum flux shear stress tensor normal to the surface gives the form drag (F_{D-1}) for a sphere

$$F_{D-1} = \pi \mu_g d(v_g - v_p) \qquad (2.7)$$

A similar evaluation for the tangential shear stress gives the friction drag (F_{D-2})

$$F_{D-2} = 2\pi \mu_g d(v_g - v_p) \qquad (2.8)$$

The total drag force (F_D) or resistance of the medium due to fluid motion relative to the particle is the sum of form and friction drag and is known as Stokes' Law:

$$F_D = 3\pi \mu_g d(v_p - v_g) \qquad (2.9)$$

If there were no relative motion, the kinetic contribution (*i.e.*, Stokes' Law) would be zero.

Total force (F) of the fluid on a spherical particle consists of the buoyant plus drag forces:

$$F = 1/6\ \pi d^3 \rho_g g + 3\pi\mu_g d(v_p - v_g) \tag{2.10}$$

The buoyant force is present whether or not motion exists.

The analytical evaluation procedures used to obtain the normal and tangential drag forces are only valid at low flow rates, where particle Reynolds number is about 0.1. Empirical corrections may be used for larger and smaller values of Re_d.

2.2 GRAVITATIONAL FORCE

2.2.1 Continuum Regime

As defined in Chapter I, Knudsen numbers of particles in this regime are < 0.1 which corresponds approximately to particles $> 1.3\ \mu m$ in diameter in air at standard conditions. As Kn approaches zero the gaseous medium acts more as a continuum to the particle. A particle in free fall in still gas acted upon only by gravitational force accelerates from rest until a terminal steady state velocity is attained. At steady state the gravitational force balances the total force of the fluid on the particle. Unless otherwise stated, conditions of low Ma number exist, and the particle is considered to be a rigid sphere. If Re_p is about 0.01, then Equation (2.10) as used with Stokes' Law can be set equal to gravitational force (F_G), where

$$F_G = 1/6\ \pi d^3 \rho_p g \tag{2.11}$$

The solution gives particle terminal settling velocity (v_s):

$$v_s - v_g = \frac{d^2 g(\rho_p - \rho_g)}{18\mu_g} \tag{2.12}$$

In cases where the gas velocity is negligible and the fluid density can be neglected compared to the particle density, this equation can be simplified to produce the more familiar expression

$$v_s = \frac{d^2 g\rho_p}{18\mu_g} \tag{2.13}$$

These equations give values of terminal settling velocity within about 1% accuracy for $0.01 < Re_p < 0.1$. This corresponds to particles from about 16-30 μm in diameter with a density of 1-2 g/cm^3 in air. At this point it becomes necessary to check by substituting v_s into Equation (1.11) as v_p to establish the value of Re_p. The error expected would increase from $< 1\%$ for Re_p values about 0.01 to about 10% at Re_p values of 1.0, as shown by Figure 1.8 for spherical particles of unit density in still air at SC of 20°C and 1 atmosphere.

For example, a particle 20 μm in diameter with a density of 1.0 g/cm^3 would have a terminal settling rate in still air of

$$v_s = \frac{(20 \times 10^{-4} \text{ cm})^2 (980 \text{ cm/sec}^2)(1.0 \text{ g/cm}^3)}{(18)(1.83 \times 10^{-4} \text{ g/cm-sec})} = 1.19 \text{ cm/sec}$$

and a particle Reynolds number of

$$Re_p = \frac{20 \times 10^{-4} \text{ cm})(1.19 \text{ cm/sec})(1.2 \times 10^{-3} \text{ g/cm}^3)}{(1.83 \times 10^{-4} \text{ g/cm-sec})} = 0.016$$

Particles this big are larger than fine particles as defined in Chapter I. However, to complete the discussion, procedures for estimating terminal settling velocity of larger particles in the Continuum Regime are now presented.

Oseen[1] accounted for nonsymmetry in the flow streamlines around a spherical particle moving through a gas, which is neglected in Stokes' derivations. This turbulence is minor for smaller particles but increases with particle size. The Oseen approximation can be used with about 1% accuracy for particles with Re_p from 0.1 to about 0.5. This procedure corrects Stokes' Equation by

$$F_D = 3\pi\mu_g d(v_p - v_g)(1 + 3/16 \, Re_p) \tag{2.14}$$

Substituting this into Equation (2.10) and combining as before with Equation (2.11) gives

$$v_p - v_g = \frac{d^2 g(\rho_p - \rho_g)}{18\mu_g(1 + 3/16 \, Re_p)} \tag{2.15}$$

Solution of this equation is by trial and error starting with assumed values of Re_p between 0.1 and 0.5.

The terminal settling velocity of even larger particles with Re_p of 3-400 can be estimated if the method of Klyachko[2] is used. This requires that the Newton Drag Equation be used in the form

$$F_D = \frac{C_D \rho_g (v_p - v_g)^2 \pi d^2}{8} \tag{2.16}$$

where the projected area of a spherical particle of $\pi d^2/4$ is used. C_D is the dimensionless drag coefficient and is often called the particle friction factor (f). It is related to particle Reynolds number by Stokes

$$C_D = \frac{24}{Re_p} \tag{2.17}$$

or by Oseen

$$C_D = \frac{24}{Re_p} + 4.5 \tag{2.18}$$

or by Klyachko

$$C_D = \frac{24}{Re_p} + \frac{4}{(Re_p)^{1/3}} \tag{2.19}$$

Use of Equation (2.16) in Equation (2.10) in place of Stokes drag expression yields the following expression in terms of C_D for gravitational effect on large particles

$$v_p - v_g = \sqrt{\frac{4d(\rho_p - \rho_g)g}{3C_D\rho_g}} \tag{2.20}$$

Any of Equations (2.17) through (2.19) can be used to replace C_D. Using Equation (2.19) the solution is best made by trial and error for Re_p from 3 to 400. A 200 μm particle of unit density falling through still air at SC would have a Re_p of about 8.8, a C_D of about 4.65 and a v_s of about 68 cm/sec.

Behavior of particles smaller in size than about 16 μm can be estimated using the Stokes Equation, but as Re_p decreases below 0.01, error in estimating v_s increases until it is about 10% at 1.6 μm. The Re_p of a 1.6 μm particle of unit density in still air at SC is about 10^{-5}. As an alternative procedure for these size particles, the Cunningham correction as discussed next can also be used.

2.2.2 Slip Flow Regime

Particles in this regime are defined in Chapter 1 as having Knudsen numbers $\leqslant 0.3$. This corresponds to particles in still air at SC with a diameter of about $\geqslant 0.4$ μm. In a sense this is considered an extension of the Continuum Regime because continuum approximations with slip flow corrections are used. Deviation from the Stokes' Equation (2.9) is noted as particle size becomes very small or as the medium gas pressure decreases. In either case the particle appears to slip through the gas molecules. Slippage of gas molecules past a surface is a deviation from continuum behavior of the medium and an indication that particles in this regime should move at a faster terminal fall velocity than predicted by continuum procedures. Hidy[3] shows that at Kn of 0.3 the free-fall terminal velocity is about 35% greater than predicted by continuum conditions and at Kn of 0.5 it is 60% greater.

One procedure is to correct the Stokes Equation by the Cunningham factor (*C*) to obtain

$$F_D = \frac{3\pi\mu_g d(v_p - v_g)}{C} \tag{2.21}$$

Use of this correction factor in Equation (2.10) gives for free-fall conditions a terminal steady-state velocity expression of

$$v_s - v_g = \frac{d^2 g(\rho_p - \rho_g)\, C}{18\mu_g} \qquad (2.22)$$

This correction can be expected to yield values of v_s within about 1% accuracy for Kn values about 0.3. This corresponds to particle diameters of about 0.36-0.80 μm in air at SC, as shown by Figure 1.8. Up to 10% error may be expected when particles as small as 0.1 μm and as large as 1.6 μm are considered (Kn = 1.3 and 0.1 respectively in air).

Numerical values of C are close to unity for particles larger than 1 μm, but the value becomes large for smaller diameters. For example, using Equation (1.22), C for 1.6 μm particles in air at SC is 1.10 and for 0.1 μm particles it is 2.86. At 55°C the values of C become 1.14 and 3.11 respectively, and at 120°C the values are 1.37 and 3.57. Many collection devices operate at temperatures such as these.

The use of Equation (2.22) can be extended to include particles with values of Kn larger than about 1.0 if C is determined using Equations (1.21) and (1.24). Note that these equations for C are useful for gaseous media other than air and for values of Kn more than 100.

2.2.3 Free Molecule Regime

This includes particles with values of Kn greater than 10. In air at SC this corresponds to particles smaller than 0.01 μm. In this regime, particles are small compared to the gas mean free molecule path, and there are few intermolecular particle-gas collisions. As a subsection under Gravitational Force, there is little of significance that can be discussed here that would be useful in air pollution control work. Particles smaller than 0.01 μm in air are likely to remain suspended, even if only temporarily, because of the gas convective motion and the presence of minute external forces. The diffusional velocity of a 0.01 μm particle in air at SC is nearly 4000 times greater than the terminal free-fall velocity of the same particle, showing that gravitational settling is almost insignificant. For smaller particles the diffusion rate increases and settling rate decreases still further. Particles this size in a working system probably never reach terminal settling conditions because of normal perturbations in the system.

Should terminal settling velocity be desired, it could be estimated using the Cunningham correction procedure noted in Equations (2.22), (1.21) and (1.24). This could give useful approximations of terminal settling velocity for particles in still air at SC with a diameter as small as 0.01 μm.

2.3 OTHER FORCES

At this point it should be apparent that the effects of other forces acting on single, rigid, spherical particles in gaseous media can be predicted in a similar manner. In some instances this may consist of only replacement of the gravitational force, Equation (2.11), by an appropriate expression for the force in consideration, resolving the particle force expression, Equation (2.10), with modifications if necessary, according to the particle size and mean free path of the gas molecules. Other cases may require inclusion of an energy balance expression in addition to equations of continuity and motion. Detailed discussions of other forces are included in Chapter 4.

2.4 PARTICLE DIFFUSION

In the treatment of the Free Molecule Regime size particles (section 2.2.3) under a gravitational force, it was noted that particle diffusion was much more significant than terminal settling for these small particles. Particles not under the influence of external forces diffuse in a random fashion called Brownian motion. Even so, each particle experiences a resultant net "unidirectional" motion, so it is appropriate to include diffusion in this chapter.

Gas molecules travel in "straight lines" and, for example, collide according to the kinetic molecular theory. The molecules are considered elastic and abruptly change speed and direction upon collision with other gas molecules. Particles, by contrast, have a much greater mass than gas molecules, yet a particle in Brownian motion in a gas has the same average kinetic energy as a gas molecule. Therefore, the particle velocity is much less than the gas mean molecular velocity which is about 450 m/sec at standard conditions. The direction and speed of particles that collide with gas molecules are affected only slightly. After numerous collisions the particle direction can be changed completely. The distance the particle moves between complete direction changes is its apparent mean path, ℓ_d.

In addition to a velocity gradient and to the internal energy, diffusion can result because of the presence of differences or gradients in concentration, pressure, external forces and temperature gradients. Diffusion due to temperature (Soret effect), pressure and external forces effects can be significant in particle diffusion and should be considered methods for removing fine particles from gaseous media. These will be covered in Chapter 4.

Diffusion flux of a given component in a mixture is proportional to the sum of the molar flux, which results from bulk flow of the fluid plus the molar flux resulting from the diffusion superimposed on the bulk flow. This second term, where a concentration gradient exists in a binary mixture, can be expressed by Fick's Law, which states that mass transport occurs because of a gradient in mass concentration. For a constant density system, neglecting the other effects, this can be written as

$$J_A = - (C_A + C_B) D_{AB} \nabla X_A \tag{2.23}$$

where J_A = molar diffusion flux of A, g moles A/(cm^2-sec)
  C_A = molar concentration of A, g moles A/cm^3 solution, etc.
  D_{AB} = mass diffusivity of component A through B, cm^2/sec
  X_A = mole fraction A

Equation (2.23) shows that A moves relative to the mixture in a direction of decreasing mole fraction A. The mass diffusivity $D_{AB} = D_{BA}$ in a binary system where equimolar counter diffusion exists. This is not true for diffusion of species A through stagnant species B. Stagnant diffusivity is more relevant for diffusion of particles through a gas.

The Nernst-Einstein Equation[4] for diffusivity of single particles through a stationary medium (D_{PM}) is

$$D_{PM} = \frac{kTv_d}{F_d} \tag{2.24}$$

where k = Boltzmann constant
   = 1.38×10^{-16} g cm^2/(sec^2 particle $^\circ$K)
  T = absolute temperature, $^\circ$K

The quantity (v_d/F_d) is the *mobility* (B) of the particle A, expressed as the steady-state velocity attained by the particle under the action of the unit diffusivity force, F_d.

For spherical particles in a stagnant gaseous medium, the diffusion force becomes Stokes' Law corrected for slip:

$$F_d = \frac{3\pi\mu_g dv_p}{C} \tag{2.25}$$

For particles in gas mediums consisting of two or more different gases, Waldmann and Schmitt[5] suggest the diffusion force should be corrected for diffusiophoresis using the correction $(1 + \sigma_{AB}X_B)/X_B$ in addition to C. The mole fraction of species B equals X_B, and the value of the factor σ_{AB} is determined empirically using

$$\sigma_{AB} = 0.95 \left(\frac{m_A - m_B}{m_A + m_B} \right) - 1.05 \left(\frac{d_A - d_B}{d_A + d_B} \right) \tag{2.26}$$

where m_A = molecular mass of species A, etc.
  d_B = molecular diameter of species B, etc.

Air may be considered as a single component with values for m_B and d_B respectively of 29 and 3.63 Å (3.63 X 10^{-4} μm). By contrast, propane has values of m_A = 44 and d_A = 5.06. This will be considered further in Chapter 4.

In case there is no tendency for the fluid to stick at the surface of the diffusing particle,[6] Equation (2.25) should be modified by multiplying by 2/3.

Combining Equations (2.24) and (2.25) gives the Stokes-Einstein Equation corrected for slip:

$$D_{PM} = \frac{CkT}{3\pi\mu_g d} \qquad (2.27)$$

This is good for particles in Slip Flow and Continuum Regime when Kn < 0.3 and Ma ≪ 1 and is also valuable for estimating diffusivity values for smaller particles, but accuracy decreases as size decreases. An expanded discussion on diffusion force is given in Section 4.5. Methods given there for obtaining F_d can be used to obtain refined values of D_{PM}.

The steady state diffusional velocity of a particle, as noted in Equation (2.24), is also called the root mean square thermal velocity, and is defined

$$v_d = \sqrt{\frac{8kT}{m_p}} \qquad (2.28)$$

where m_p is the mass of the particle [see also Equation (2.33)]; the Cunningham factor should be included under the square root sign where needed. The mean square velocity ($\overline{G}^2$) is 3/8 the square of v_d and is

$$\overline{G}^2 = \frac{3kT}{m_p} \qquad (2.29)$$

while the mean square displacement along one axis in rectangular coordinates would be

$$\overline{G}_x{}^2 = (1/3)\,\overline{G}^2 \qquad (2.30)$$

Note that for small spherical particles the mean square velocity can also be expressed as

$$\overline{G}^2 = \frac{3D_{PM}}{\tau} \qquad (2.31)$$

where τ is a characteristic or relaxation time defined as

$$\tau = \frac{d^2 \rho_p}{18\mu_g} \qquad (2.32)$$

and in this instance t ≫ τ. That is to say that steady state conditions exist. Significance of relaxation time is discussed in Chapter 3.

A better approximation for root mean square thermal velocity for fine particles would be to combine Equations (2.27) and (2.31) to obtain

$$v_d = \sqrt{\frac{8 D_{PM}}{\tau}} \qquad (2.33)$$

The apparent mean free path of a particle in a given direction as a result of diffusion (ℓ_d) is

$$\ell_d = v_d \tau \qquad (2.34)$$

where again $t \gg \tau$. This is equally dependent on both particle size and particle density.

The particle has a random motion, and by probability theory the net distance traveled by a particle in a given time can be estimated. The mean displacement due to diffusion is

$$\overline{\Delta X}_d = \sqrt{\frac{4 D_{PM} t}{\pi}} \qquad (2.35)$$

The thermal mobility (B), as noted earlier, is equal to (v_p/F_d) where here, the particle velocity v_p is the mean diffusional velocity v_d. Using Equation (2.25) this can also be expressed as

$$B = \frac{C}{3\pi\mu_g d} \qquad (2.36)$$

This is also called diffusional or mechanical mobility. Values of B in units of sec/g are plotted against diameter in Figure 2.1.

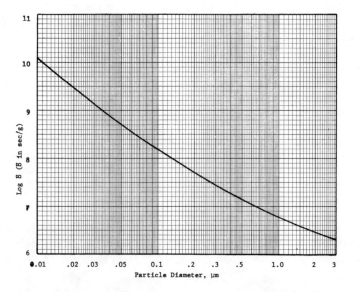

Figure 2.1 Thermal mobility, B, of spherical particles in air at SC.

Table 2.1 shows the expected behavior of spherical particles with a density of 1.0 g/cm^3 in still air at SC using μ_g = 1.83 X 10^{-4} g/(cm-sec). Accuracy is good for the size particle listed. The calculated particle mechanical mobility values reported in Table 2.1 agree with the mobility of oil drops given by Fuchs[6] and based on the data of Millikan.[7] Agreement is such that the noted mobility ranges from 1.02% below Fuchs' value for 1.6 μm particles, to only 0.16% above for 0.01 μm particles. The diffusivities listed agree with the reported values to within 0.75%. Corrections for particles of densities other than 1.0 g/cm^3 can be made by multiplying the listed values by density to the exponent noted at the right of the table. Values for other temperatures can be obtained by using appropriate temperatures and medium viscosities, where necessary.

2.5 AERODYNAMIC DIAMETER

Aerodynamic diameter (d_a) is defined in Section 1.3 as the diameter of a sphere of unit density that attains the same terminal (steady state) settling velocity (v_s) at low Reynolds number in still air as the actual particle under consideration. The particle need not be in free fall to apply this correlation. It is more likely that the particle is moving in a gaseous medium and the inflight or aerodynamic nature is the property observed. It is desirable to be able to correlate this to a true size and as pointed out by Cooper,[8] aerodynamic diameter may or may not equal true diameter. Aerodynamic diameter is commonly used rather than geometric diameter when discussing collection efficiency as a function of particle size for specific devices. Inertial impaction and gravitational forces are proportional to the particle mass and in these cases aerodynamic diameter is the particle size that governs particle motion.

Terminal settling velocity of a spherical particle in still air can be found using the procedures discussed earlier in this chapter and summarized in Table 2.2. According to the definition given above for aerodynamic diameter, the terminal settling velocity, (for example) when Re > 3 X 10^{-7}, can be found using Equation (2.22) in the form

$$v_s = \frac{d_a^2 \, g \, \rho_o \, C_a}{18 \, \mu_g} \qquad (2.37)$$

where ρ_o = unit density, 1 g/cm^3

C_a = Cunningham correction applied using d_a

Terminal settling velocity in terms of aerodynamic diameter can be expressed for other size particles using the appropriate equation from Table 2.2 and converted in the same manner as is used to obtain Equation (2.37). Particles

Table 2.1   Calculated Diffusional Behavior of Spherical Particles with a Density of 1 g/cm³ in Still Air at Standard Conditions

Diffusional Properties	Particle Diameter (μm)						For Other Densities, Correct Using
	1.6	1.0	0.50	0.10	0.05	0.01	
Cunningham factor — Equation (1.22) $C = 1 + \dfrac{(2\times10^{-4})\,T}{d_p}\left\{2.79 + 0.894\exp\left(-\dfrac{2.47\times10^3\,d_p}{T}\right)\right\}$	1.10	1.16	1.33	2.86	4.96	22.2	ρ_p^0
Diffusivity, cm²/sec — Equation (2.27) $D_{PM} = \dfrac{C\,k\,T}{3\pi\,\mu_g\,d}$	1.61×10^{-7}	2.72×10^{-7}	6.24×10^{-7}	6.71×10^{-6}	2.33×10^{-5}	5.20×10^{-4}	ρ_p^0
Relaxation time, sec — Equation (2.32)[a] $\tau = \dfrac{d^2\,\rho_p\,C}{18\,\mu_g}$	8.55×10^{-6}	3.53×10^{-6}	1.01×10^{-6}	8.69×10^{-8}	3.77×10^{-8}	6.75×10^{-9}	ρ_p^1
Root mean square thermal velocity, cm/sec — Equation (2.33) $v_d = \sqrt{\dfrac{8D_{PM}}{\tau}}$	0.39	0.79	2.22	24.9	70.3	785	$\rho_p^{-1/2}$
Particle apparent mean free path, cm — Equation (2.34) $\ell_d = v_d\,\tau$	3.32×10^{-6}	2.79×10^{-6}	2.24×10^{-6}	2.16×10^{-6}	2.65×10^{-6}	5.30×10^{-6}	$\rho_p^{-1/2}$
Mean displacement due to diffusion, cm in 1 sec — Equation (2.35) $\overline{\Delta X_d} = \sqrt{\dfrac{4D_{PM}\,t}{\pi}}$	4.53×10^{-4}	5.89×10^{-4}	8.91×10^{-4}	2.92×10^{-3}	5.45×10^{-3}	2.57×10^{-2}	ρ_p^0
Thermal mobility, sec/g — Equation (2.36) $B = \dfrac{C}{3\pi\,\mu_g\,d}$	3.99×10^6	6.73×10^6	1.54×10^7	1.66×10^8	5.75×10^8	1.29×10^{10}	ρ_p^0

[a]See Section 2.1.

Table 2.2 Terminal Settling Velocities of Spherical Particles in Still Air

Regime		Equation	Approximate Particle Reynolds Number in Air at SC
Continuum	(2.20)	$v_s = \left[\dfrac{4d\, \rho_p\, g}{3\, C_D\, \rho_g} \right]^{1/2}$	3-400
	(2.15)	$v_s = \dfrac{d^2\, g\, \rho_p}{18\, \mu_g\, (1 + 3/16\, Re_p)}$	0.1-0.5
	(2.13)	$v_s = \dfrac{d^2\, g\, \rho_p}{18\, \mu_g}$	0.01-0.1
Slip Flow	(2.22)	$v_s = \dfrac{d^2\, g\, \rho_p\, C}{18\, \mu_g}$	$> 3 \times 10^{-7}$
Free Molecule	(2.22)	$v_s \cong \dfrac{d^2\, g\, \rho_p\, C}{18\, \mu_g}$	$> 3 \times 10^{-13}$

approximately in the range covered by the Cunningham corrected Stokes Equation are highly significant in pollution control studies, so Equation (2.22) is used as the example for the rest of this discussion.

Equations (2.37) and (2.22) can be set equal to obtain a value of true diameter in terms of aerodynamic diameter or vice-versa:

$$d^2\, \rho_p\, C = d_a^2\, C_a \qquad (2.38)$$

Solution of this equation is possible using the simplified Equation (1.23) for the appropriate Cunningham factor and the quadratic equation; however, often it is easier to solve by trial and error. For example, obtain d knowing aerodynamic diameter (d_a) by using

$$d = d_a \sqrt{\frac{C_a}{C\, \rho_p}}$$

Equations (1.21-1.23) can be used to obtain C_a from d_a, and with the value of true density (ρ_p) everything on the right side is known except C. As a first approximation, $C = C_a$ could be assumed. When the value of d for this C is obtained, a new C can now be calculated and used to produce the corrected d. A third iteration or more may be needed depending on the accuracy desired. Remember,

$$d_a = d \text{ when } \rho_p = 1.0 \text{ g/cm}^3$$
$$d_a > d \text{ when } \rho_p > 1.0$$
$$d_a < d \text{ when } \rho_p < 1.0$$

Aerodynamic diameters can be substantially different from true diameters. A particle in air at SC with an aerodynamic size of 0.5 μm would have a true diameter of 0.34 μm at a specific gravity of 2, and 0.74 μm at 0.5.

2.6 SHAPE EFFECTS

Section 1.3 provides a procedure that enables us to account for non-spherical particles as if they were spherical. If we deal strictly with fine particles as defined (*i.e.,* $d_a < 3 \ \mu$m), the effects of shape variations become less significant than for larger particles. The particle aspect ratio (length of major axis to length of minor axis) may still be large, but as the major axis increases the particle aerodynamic size increases, and the particle may no longer be categorized as a fine particle. Fine particles are quite likely to be produced by condensation, which often results in essentially spherical shapes. Other particles produced by atmospheric reactions are likely to have a fairly symmetrical crystalline shape, which makes them nearly spherical. Even boiler fly ash fine particles appear essentially as rough spheres as seen at 11,500 magnification in Figure 2.2. These particles were collected aerody-namically from a boiler flue gas stream using a silver foil in an inertial im-pactor. This electron microscope picture shows the particles, ranging from about 0.1-1 μm in diameter, as bright spots collected on one segment of the porous silver foil. These are not the hollow cenospheres formed by con-densing molten slag; those are much larger.

Many studies conducted to measure the behavior of particles in gaseous media use oil drops in air at SC. This is convenient because oil can be easily atomized to form desired sizes of spherical particles. These droplets maintain their size because the oil has a low volatility, is hydrophobic and is not soluble in water. In addition, oil has a high viscosity (compared to water), so the drag effects that cause internal circulation in less viscous particles are less significant for the oil drops.

The effect of internal circulation in a low-viscosity liquid droplet is to reduce the medium drag resistance. This is shown schematically in Figure 2.3 for a droplet in free fall and can be adapted to movement resulting from any force. The extent of the effect on drag force on a droplet in the Continuum Regime is given as a modification of Stokes' Law,[9] Equation (2.9):

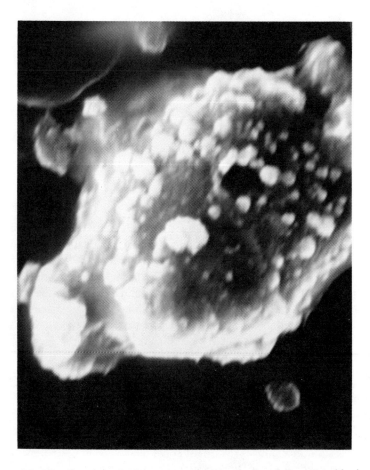

Figure 2.2   Fine fly ash particulate matter on a porous silver foil, captured by inertial impaction. (Electron microscope picture at 11,500 magnification.)

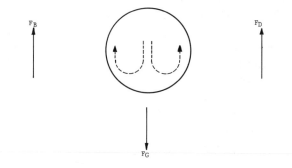

Figure 2.3   Liquid droplet in free fall showing internal circulation in a low-viscosity liquid.

$$F_D = 3\pi \mu_g \, d \, V_\infty \frac{1 + (2/3)(\mu_g/\mu_p)}{1 + \mu_g/\mu_p} \tag{2.39}$$

where V_∞ is the velocity of the medium far from the droplet relative to the velocity of the droplet. This correction is completely negligible for oil and amounts to 0.58% for water in air at SC.

2.7 PARTICULATE CLOUDS

Groups of fine particles in a cloud, both in the atmosphere and in the confinement of emission ducts and collection devices, have characteristic properties that make them behave differently than single particles. This effect due to the presence of a large number of particles, causing them to move faster or slower than single particles is called hydrodynamic interaction. This section deals with large numbers of particles. Interaction effects of one particle on another are discussed in Chapter 5. Natural clouds are formed when humid air condenses to form small droplets of suspended water. Meteorologists consider mists to be 1-5 μm droplets and clouds as 5-200 μm droplets with drizzle as 200-500 μm and rain as 500-8000 μm. For our purposes, all can be considered clouds of particles although most natural cloud water droplets are < 40 μm. Frequently, raindrops, although falling under the influence of gravity, have a net upward velocity because the entire cloud is rising at a rate faster than the drops are falling. Clouds may also be groups of condensed hydrocarbon droplets, groups of solid particles or any combination of these and water droplets. Natural clouds average about 1 g/m^3 water vapor and exhibit temperature and pressure variations from their surroundings.

Clouds are also produced by thermal and mechanical action. Forest fires, for example, frequently produce clouds above the fire when the water vapor produced during combustion rises to a point of condensation. Large power plants and large concentrations of people operating cars and using combustion heating also can produce clouds, as can the combination mechanical-thermal action of explosive detonations. Atomization is a favored method for producing clouds in collection devices and can also be used to produce atmospheric clouds.

Clouds of particles with a density greater than that of the gaseous medium exhibit a hydrostatic force. This can be physically observed by looking at the surface of fluidized beds and noting the wave-like appearance that makes it resemble a liquid. This force resists disturbances of the medium and tries to maintain the identify of the cloud. Figure 2.4 shows individual droplets of water atomized to < 10 μm in diameter moving down the diverger of a venturi scrubber in a cloud formation.[10] This is a two-dimensional view

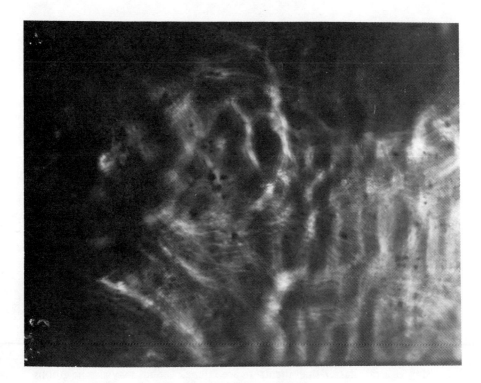

Figure 2.4 Cloud-type atomized water in venturi diverger showing clouds of approximately 500 μm effective diameter. (11-fold magnification, 7.4 ℓ water/m^3 air @ SC; motion direction →)

of the scrubber with flow from left to right. The travel of these clouds was observed from the throat to a distance of 23 cm beyond the throat, and the clouds retained their identity extremely well. In the same study, it was shown that effective cloud diameters are linearly related to scrubber atomizing throat velocity in this system as shown in Figure 2.5. This was verified using both physical measurements (stop-action photography) and drag coefficient-Reynolds number correlations.

Clouds of particles in the atmosphere and clouds confined in systems that are large enough to make surface effects negligible have a relatively flat top surface when the medium motion is not too violent. This exists because the hydrostatic force of the cloud, like a surface tension effect, attempts to counteract the gravitational force on the particles that are heavier than the gaseous medium. Clouds acted upon by forces strong enough to distort the cloud will have shapes corresponding to the medium forces such as those in Figure 2.4 and those commonly seen in atmospheric clouds.

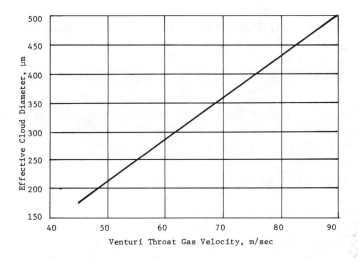

Figure 2.5 Effective diameter of atomized clouds as a function of venturi velocity using 1.6 mm water inlet nozzles at throat.

Fuchs[6] shows that clouds as initially formed move as a unity. Later, as the cloud is dispersed by air currents and diffusion, the volume increases and the hydrostatic force is dissipated so that the gas blows through the cloud and it ceases to exist. The character of motion (M) is given as

$$M = 0.27 \, \pi \, N \, D_c^2 \, d'$$ (2.40)

where N = number of particles in cloud per cm^3
D_c = effective cloud diameter, cm
d' = particle diameter, cm

When $M \gg 1$, the cloud moves as an entity. The cloud can be blown through at lower values of M. For comparison, a natural cloud contains about 1 g/m^3 of liquid drops. The cloud in the scrubber shown in Figure 2.2 has 5.5 gal H_2O/1000 ft^3, which equates to about 500 g/m^3 of water assuming 70% of the water is in the atomized phase. If the clouds are made up of particles with an average diameter of 5 μm, the number concentration for the natural cloud (N_N) is 1.5 X 10^4 particles/cm^3 and for the scrubber cloud (N_S) is 7.6 X 10^6 particles/cm^3. (These scrubber droplets would occupy a volume of about 5 X 10^{-4} cm^3.) The resultant M for a ½ km natural cloud and a 500 μm scrubber cloud are 1.6 X 10^{10} and 8, respectively. The natural cloud would be expected to remain intact for a considerable period of time while the scrubber cloud should be blown through relatively soon. (Note: hold-up time in the scrubber diverger at these conditions is only about 0.01 sec.)

Very concentrated clouds under the influence of gravity have settling rates for the entire system different from the velocity of a particle of the same size, determined by

$$V \cong v_s(1 - \phi)^{4.65} \tag{2.41}$$

where V is the settling velocity of the cloud, v_s is single particle settling rate and ϕ is the fraction of volume occupied by the particles. This shows that even the scrubbing cloud composed of 5 μm droplets used in the previous example, if it were free to settle, would have a settling rate essentially the same as that of the free fall rate of a 500 μm water particle.

2.8 EFFECTS OF ADJACENT SURFACES

In atmospheric studies and in large-scale systems, the effects of surfaces are usually neglected; however, the presence of surfaces can be significant and will be discussed. In normal viscous flow of a fluid through a stationary container, the fluid is considered to be stationary or nearly stationary at the surface because of the drag effects from the surfaces. In nonviscous (*i.e.*, potential) flow, viscosity is assumed to be zero, and all the fluid moves at the same velocity. The greater the fluid viscosity, the greater the difference in flow between fluid at the walls and fluid in the center of the system. A normal (Newtonian) fluid in steady state laminar flow in a cylindrical tube of radius R has an axial velocity v_z at any radius r of

$$v_z = v_{z\,max} \left[1 - (r/R)^2 \right] \tag{2.42}$$

where $v_{z\,max}$ is the maximum velocity that occurs at r = 0 and has the value

$$v_{z\,max} = \frac{[(P - \rho_g gh)_o - (P - \rho_g gh)_L]\ R^2}{4\,\mu_g\ L} \tag{2.43}$$

where P = pressure
      h = elevation above a reference plane
      L = distance from starting location, o

Subscripts o and L represent conditions at start and location L, respectively. These equations apply to incompressible fluids, and air is considered incompressible near normal conditions. Therefore, for a gaseous fluid in a horizontal system the terms in brackets in Equation (2.43) reduce to the pressure difference. In a circular tube the bulk average fluid velocity, v_{av}, with laminar flow is

$$v_{av} = \tfrac{1}{2}\,v_{z\,max} \tag{2.44}$$

and with turbulent flow is

$$v_{av} \cong 4/5 \ v_{z \ max} \qquad (2.45)$$

Laminar flow along flat surfaces gives

$$v_{av} = 2/3 \ v_{max} \qquad (2.46)$$

The force of the fluid in laminar flow on the wetted surface of a cylindrical pipe is

$$F_D = \pi R^2 \ [(P_O - P_L) + L \ \rho_g \ g] \qquad (2.47)$$

The fluid medium exerts drag forces on particles in the medium as well as the adjacent surfaces. This is true whether the fluid, particles, surfaces or any combination have motion relative to the others. Particle movement, for example, in a stationary gas, results in gas movement. The gas movement could be slowed by the presence of a stationary surface, which propagates the effect of slowing the particle because of the presence of the surface. Behavior of Continuum Regime spherical particles near surfaces can be accounted for by a modification of the Stokes Equation (2.9), as given by Happel and Brenner:[11]

$$F_D = \frac{3 \ \pi \ \mu_g \ d \ v\infty}{(1 - C_1) \ (d/2\ell) + 0.215 \ (d/2\ell)^3} \qquad (2.48)$$

where ℓ is the distance from the center of the spherical particle to the surface and C_1 is a constant. This work shows that the surface effects only become significant when the particle is within about 5 diameters distance from the wall. The surface effect becomes increasingly significant as the particle approaches the wall. Values of C_1 for spherical particles are 2.104 for motion along the axis of a spherical container, 0.563 for motion parallel to a flat plane, and 1.125 for motion perpendicular to a flat plane.[11] Experiments predict that "effects" on clouds of particles may be increased over that indicated by Equation (2.48) by the approximation $(1 + \phi)$, where ϕ is the fraction of volume occupied by the particles in the cloud.

Surfaces can influence particle behavior because of adhesive, electrostatic, ionic and other forces. These effects can be much greater than those of isolated surfaces and will be covered in Chapter 4. Effects of targets placed in a flowing fluid to collect particles and of effects of particle interaction are covered in Chapter 5.

REFERENCES

1. Oseen, C. *Neure Methoden und Ergebnisse in der Hydrodynamik* (Leipzig: Akademische Verlag, 1927).

2. Klyachko, L. *Otopl. i. Ventil.* No. 4 (1934).
3. Hidy, G. M. and R. J. Brock. *The Dynamics of Aerocolloidal Systems* (London: Pergamon Press, 1970).
4. Daniels, F. and R. A. Alberty. *Physical Chemistry.* (New York: Wiley, 1955), p. 650.
5. Waldmann, L. and K. Schmitt. *Aerosol Science.* C. N. Davis, Ed. (New York: Academic Press, 1966).
6. Fuchs, N. A. *Mechanics of Aerosols* (London: Pergamon Press, 1964).
7. Millikan, R. *Phys. Rev.* 22(1) (1923).
8. Cooper, D. W., R. Wang and D. P. Anderson. "Evaluation of 8 Novel Fine Particle Collection Devices," EPA-600/2-76-035 (February 1976).
9. Rybcznski, W. *Bull. Acad. Sci. Cracovie A.* 40 (1911).
10. Hesketh, H. E. "Atomization and Cloud Behavior in Venturi Scrubbing," *J. Air Poll. Control Assoc.* 23(7):600 (1973).
11. Happel, J. and H. Brenner. *Low Reynolds Number Hydrodynamics* (Englewood Cliffs, New Jersey: Prentice-Hall, 1965).

CHAPTER 3

NONSTEADY-STATE MOTION OF PARTICLES

Acceleration and deceleration of particles are extremely important to the control of fine particles. It is necessary to change the velocities of moving particles to remove them from a moving gas stream so deceleration will occur whenever a particle is captured. In many control devices, the particles are accelerated to high velocities to enhance capture of fine particles. During any period of velocity change, the particle is in unsteady state motion, and behavior cannot be predicted directly by the methods used for steady state behavior. The time that a particle is in an unsteady state condition is usually small compared with the time in steady state motion, but the significance of this condition should not be underestimated. In cases where a constant force is applied to a particle over a relatively long period of time, we can neglect the unsteady state acceleration or deceleration and assume that the particle always moved at the steady-state condition. This is called quasi-stationary and is assumed in the Chapter 2 discussions.

3.1 CHARACTERISTIC TIME

Relaxation time, τ, is introduced in Section 2.4 in the discussion of particle diffusion. This time can be thought of as being the time physically during which *most* of the motion change occurs when a force is applied to a particle starting from a steady state condition. The initial steady state is either an at-rest or in-motion condition. At the point where time $t = \tau$, a particle being accelerated reaches approximately two-thirds the terminal velocity and a particle being decelerated reaches about one-third the initial velocity. It is apparent that great change in particle motion occurs when $t < \tau$, and little change occurs when $t > \tau$. Therefore, τ is the time during which most change of motion energy is dissipated.

During steady state motion the force acting on the particle equals the sum of buoyant and drag forces. For the conditions under which Stokes' Law, Equation (2.9), applies and $v_p \gg v_g$, $m_p \gg m_g$ and the applied force is significantly greater than gravitational force, the equality is obtained:

$$m_p \, a = 3\pi \, \mu_g \, d \, v_p \qquad (3.1)$$

The force acting on the particle is expressed as mass times acceleration (a). Resolving this expression in units of time gives a relaxation time of

$$\tau = \frac{v_p}{a} = \frac{m_p}{3\pi \, \mu_g \, d} \qquad (3.2)$$

which for a spherical particle gives the Equation (2.32):

$$\tau = \frac{d^2 \, \rho_p}{18 \, \mu_g} \qquad (2.32)$$

When gravity is the force applied to a particle at rest, $v_s = \tau g$.

Note from this that application of other forces such as electrostatic and magnetic will produce various terminal velocities but a given type and size particle will have the same relaxation time.

Relaxation times for various size and shape particles can be obtained by applying the appropriate factors as discussed in Chapters 1 and 2. For example, the value of τ for particles in the Cunningham correction region is obtained by multiplying the right side of Equation (3.2) by C or by using actual v_p/a. Values of τ for particles in air at SC are given in Table 2.1.

Other characteristic times are useful in depicting specific properties of collection devices and are discussed in Chapter 5.

3.2 NONSTEADY-STATE BEHAVIOR

3.2.1 Velocity

The simple-steady state force balance of external force $F_E = F_D + F_B$ does not apply during acceleration and deceleration. Instead, the resultant sum of forces (F_R) acting on a particle in a uniform medium becomes

$$F_R = F_E - F_B - F_D - F_{MA} \qquad (3.3)$$

The symbol F_{MA} represents added resistance forces on the particle due to change in velocity of the medium plus the added resistance caused when the particle changes the velocity of the adjacent medium as the particle changes velocity. Where Ma $\ll$ 1 the values of all forces represented by F_{MA} are negligible compared with the other forces. All external forces that can change its motion when applied to the particle must be included in F_E.

In the Presence of External Forces:

Only gravitational force (F_G) has been covered thus far, so the following treatment uses this as the example. The presence of other forces can be handled in a similar manner. It is assumed in the following derivation that $F_E = F_G = m_p g$ and that the density of gas is not significant compared to the particle density. With the assumptions given, Equation (3.3) can be written for particles that follow Stokes' Law and when $v_p \gg v_g$:

$$m_p \frac{dv_p}{dt} = m_p g - 3\pi \mu_g d \, v_p \qquad (3.4)$$

In terms of relaxation time this is

$$\frac{dv_p}{dt} + \frac{v_p}{\tau} = g \qquad (3.5)$$

This is a first-order linear differential equation of the type

$$\frac{dy}{dt} + f(t) \, y = g(t) \qquad (3.6)$$

where y is a dependent variable and f and g are arbitrary functions of the independent variable t. The dependent variable solution can be obtained:

$$y = e^{-\int f(t)dt} [\int g(t) e^{+\int f(t)dt} dt + c] \qquad (3.7)$$

where e is the natural logarithm base and c is the constant of integration.
The solution of Equation (3.5) for any Stokes size particle in free fall (*i.e.*, terminal Re_p is > 0.01 and < 0.1) starting from rest is

$$v_p = \tau g(1 - e^{-t/\tau}) = v_s (1 - e^{-t/\tau}) \qquad (3.8)$$

Section 2.2.1 shows that $v_s = \tau g$ for Stokes particles and is τCg for Cunningham particles. Equation (3.8) shows that when $t = \tau$, $v_p = 0.632 \, v_s$ and when $t \gg \tau$, $v_p = v_s$ for particles in the Stokes Regime.
Velocity of particles that fall in the Cunningham correction size range can be determined by applying the appropriate correction. Maintaining all other assumptions and remembering that τ must be correct in terms of either observed v_p/a or C, this gives

$$v_p = \tau Cg(1 - e^{-t/\tau C}) = v_s(1 - e^{-t/\tau C}) \qquad (3.9)$$

It should be noted that τ is a function of C and Stokes' Law was used assuming $v_p \gg v_g$. Smaller particles in free fall move more slowly, and if the gaseous medium is in motion this assumption should be reviewed to ensure it is applicable.

Nonsteady-state velocities of larger particles can be estimated using Newton's Equation (2.16) in place of Stokes' Law. This gives, in place of Equation (3.4),

$$m_p \frac{dv_p}{dt} = m_p g - \frac{C_D \, \rho_g \, v_p^2 \, \pi d^2}{8} \qquad (3.10)$$

when $v_p \gg v_g$. The drag coefficient C_D is a function of Re_p as discussed in Section 2.2.1. Re_p in turn is a function of v_p, and Equation (1.11) can be rearranged and differentiated at constant gas viscosity and density to obtain

$$\frac{dv_p}{dt} = \frac{\mu_g}{d\rho_p} \frac{dRe_p}{dt} \qquad (3.11)$$

Fuchs'[1] solution of Equations (3.10) and (3.11) with appropriate integration limits yields, for a spherical particle falling from rest,

$$\int\limits_{Re_p = 0}^{Re_{pf}} \frac{dRe_p}{C_1 - C_D Re_p^2} = \frac{3\mu_g \, t}{4d^2 \, \rho_p} \qquad (3.12)$$

where Re_{pf} = final particle Reynolds number
       t = time from start of free fall
      C_1 = $4g \, \rho_p \rho_g \, d^3 / 3\mu_g^2$

If the particle is being accelerated by gravity but is already in motion at the starting time, the initial Re_p must be used, and t is replaced by final and initial times $(t_f - t_o)$.

The particle velocity is obtained by solving the equation graphically to obtain a final value of Re_p. Knowing Re_p, v_p can be obtained. The area under the curve $1/(C_1 - C_D Re_p^2)$ versus Re_p is equal to the integral given in the left side of Equation (3.12). The trial and error solution is complete at a value of Re_p that makes this area equal to the value obtained by the right side. The final Re_p must be no more than steady state Re_p if all work is performed accurately. Figure 3.1 is given to simplify obtaining values of C_D - Re_p. These data are obtained using the applicable relations given in Section 2.2.1.

No External Forces Present:

When no external forces are present the F_E term in Equation (3.3) is zero, and in subsequent equations such as numbers (3.4) and (3.10) there is no middle term. This situation can exist, for example, when a particle is decelerated from steady-state motion considering only motion in a horizontal direction. Values of this stopping velocity are obtained from the appropriate equations. For a Stokes' particle the velocity is

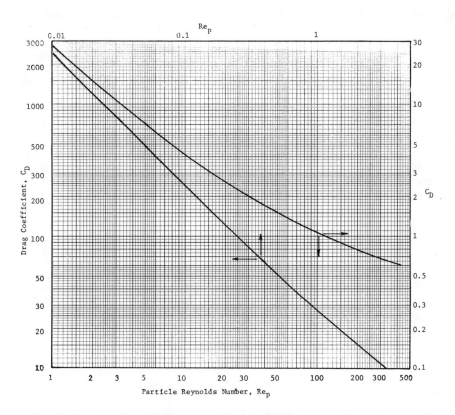

Figure 3.1 Drag coefficient as a function of particle Reynolds number.

$$v_p = v_i e^{-t/\tau} \qquad (3.13)$$

where v_i is initial velocity and t is time since external force was removed. When $t = \tau$ the final particle velocity equals 0.368 times the initial velocity and at large t, the final velocity is zero.

A Cunningham particle stopping velocity can be estimated by

$$v_p = v_i e^{-t/\tau C} \qquad (3.14)$$

Stopping velocities of particles larger than those following Stokes' Law can be estimated by a procedure analogous to that used to produce Equation (2.12). Solution in terms of Re_p gives

$$\int_{Re_{po}}^{Re_p = 0} \frac{d\, Re_p}{C_D\, Re_p^2} = -\frac{3\, \mu_g\, t}{4\, d^2\, \rho_p} \tag{3.15}$$

where Re_{po} is the original Re_p and t is the stopping time. If a final partial velocity is desired, the correct final Re_p is used (instead of 0) and $(t_f - t_o)$ is used in place of t. As before, a graphical solution is indicated where $1/C_D\, Re_p^2$ is plotted against Re_p.

3.2.2 Distance Traveled

In the Presence of External Forces:

The integral of the appropriate unsteady-state velocity equation with respect to time gives the distance traveled, x, by the particle. The equations prepared in the previous section, where gravity is the only external force, are used here. Integration of the right side of Equation (3.8) gives

$$x = v_s t - v_p \tau \tag{3.16}$$

for a particle in the Stokes Region in free fall starting from rest with no other external forces. When $t \gg \tau$ the second term is negligible and quasi-stationary motion exists.

For Cunningham correction particles the distance is obtained by integrating Equation (3.9):

$$x = v_s t - v_s \tau C (1 - e^{-t/\tau C}) \tag{3.17a}$$

$$= v_s t - v_p \tau C \tag{3.17b}$$

$$= \tau C (gt - v_p) \tag{3.17c}$$

Equation (3.17) is given in three forms as a reminder that in this case v_p is obtained from Equation (3.9), v_s is a function of τ and both v_s and τ are functions of C.

Nonsteady-state distance traveled in free fall from rest by particles larger than those covered by Stokes' Law can be estimated using a continuation of the procedure for Equation (3.12):

$$\int_{Re_p = 0}^{Re_{pf}} \frac{Re_p\, dRe_p}{C_1 - C_D\, Re_p^2} = \frac{3\, \rho_g\, x}{4\, d\, \rho_p} \tag{3.18}$$

Units are as defined for Equation (3.12). If the particle is not initially at rest, an appropriate initial Re_p is used in place of zero. Graphical solution can be made by plotting $Re_p/(C_1 - C_D\, Re_p^2)$ versus Re_p and obtaining area under the curve.

No External Forces Present:

Distance traveled in the absence of external forces is stopping distance (x_s), which is a very important parameter in the removal of particulate pollutants. **Distance** traveled in time t by a Stokes particle is obtained from integration of Equation (3.13):

$$x = v_i \, \tau \, (1 - e^{-t/\tau})$$ (3.19)

At large values of t, the stopping distance is

$$x_s = v_i \, \tau$$ (3.20)

Cunningham particle distance is estimated using

$$x = v_i \, \tau C (1 - e^{-t/\tau C})$$ (3.21)

Particles larger than Stokes have a stopping distance found by

$$\int_{Re_{po}}^{Re_p = 0} \frac{d \, Re_p}{C_D \, Re_p} = - \frac{3 \, \rho_g \, x_s}{4 \, d \, \rho_p}$$ (3.22)

If the final Re_p is not zero, the appropriate value must be used, and distance traveled, instead of x_s, is obtained.

3.2.3 Acceleration

This is the derivative of velocity with respect to time and can be found from the velocity equations. The expressions are for a continuous application or removal of external forces.

In the Presence of External Forces:

When a constant gravitational force is applied, particle acceleration (*a*) becomes, for Stokes particles:

$$a = g e^{-t/\tau}$$ (3.23)

and for Cunningham particles:

$$a = g e^{-t/\tau C}$$ (3.24)

No External Forces Present:

Deceleration resulting from the removal of external forces becomes, for Stokes particles:

$$a = -\frac{v_i}{\tau} e^{-t/\tau} \tag{3.25}$$

and for Cunningham particles:

$$a = -\frac{v_i}{\tau C} e^{-t/\tau C} \tag{3.26}$$

3.3 GRAVITATIONAL SETTLING OF PARTICLES

3.3.1 Laminar Flow

When Reynolds flow number of a gas in a system with smooth walls is less than about 1200, the flow can be considered laminar. Particles moving with a gas in horizontal flow will settle from the gas because of gravitational force. The velocity of the particles can be resolved vectorially using the forward motion component and a vertical component obtained by procedures presented in Section 2.2. Gas-particle slippage is considered negligible and particle forward motion component (v_x) is usually considered equal to gas forward velocity (u_x) at any given distance from the plate surface, when the horizontal flow is relatively uniform. Variations in flow near the surface of the plate and at the entrance and exit are neglected.

Horizontal velocity distribution in an incompressible viscous fluid in horizontal laminar flow between two flat plates (rectangular duct) can be shown by a momentum balance[2] to be

$$u_x = 6 \left(\frac{z}{h} - \frac{z^2}{h^2} \right) \bar{U} \tag{3.27}$$

where z is the distance a gas stream (or particle) is from the surface of the lower plate and h is the distance between plates. This shows that the average bulk velocity, $\bar{U}$, in the duct is equal to two-thirds the maximum velocity ($v_{x,max}$ or $u_{x,max}$). The maximum velocity occurs at the center between the plates.

Fluids flowing at constant density and viscosity can have velocity components expressed in terms of a stream function (Ψ). Physically, lines designated by constant Ψ are streamlines. In two-dimensional flow (x,z) the stream functions for the gas (and the particles) are defined:

$$u_x = \frac{\partial \Psi}{\partial z} \quad \text{and} \quad u_z = -\frac{\partial \Psi}{\partial x} \tag{3.28a,b}$$

It is also true that

$$u_x = \frac{dx}{dt} \quad \text{and} \quad u_z = \frac{dz}{dt} + v_p \tag{3.29a,b}$$

In systems where the distance between plates is far enough so that quasi-stationary free fall can be assumed, $v_p = v_s$. Fuchs[1] shows that these expressions can be used to develop expressions for the fraction of any specific size particle, ϵ_i, that will settle out of the gas during the travel of the particle between two flat horizontal plates:

$$\epsilon_i = \frac{v_s\, L}{\bar{U}\, h} \tag{3.30}$$

where L is the length of the plates. A critical plate length, L_{cr}, is required to permit complete settling of specific size particles:

$$L_{cr} = \frac{h\,\bar{U}}{v_s} \tag{3.31}$$

Overall settling efficiency can be obtained by calculating individual efficiencies using Equation (3.30) for the size range of particles in the system. A plot, for example, of cumulative percent undersize versus individual efficiency for each particle size, as described in Hesketh,[3] will give overall efficiency. Efficiencies of geometrically similar systems can be related by the ratio of terminal settling velocities to bulk flow rate of the medium for specific size particles.

Laminar flow in a horizontal cylindrical tube can be examined in a similar fashion using the same assumptions and conditions, noting in cylindrical coordinates that the horizontal velocity distribution through the tube is

$$u_x = 2\bar{U}\left[1 - \left(\frac{r}{R}\right)^2\right] \tag{3.32}$$

where r is any distance along radius of the tube and R is the radius of the tube. This shows that the gas average of bulk velocity, $\bar{U}$, is one-half the maximum. Thomas[4] notes that the fractional efficiency for a given size particle in this case, where only gravitational force acts to remove the particle, is

$$\epsilon_i = \frac{2}{\pi}\left(2C_2\sqrt{1 - C_2^{2/3}} + \sin^{-1} C_2^{1/3} - C_2^{1/3}\sqrt{1 - C_2^{2/3}}\right) \tag{3.33}$$

where $C_2 = (3v_s\, L)/(8\, R\, \bar{U})$ and arc sin is from 0 to $\pi/2$ radians.

The critical length to allow all the particles of a given size to settle by gravity from the tube is

$$L_{cr} = \frac{8\, R\, \bar{U}}{3\, v_s} \tag{3.34}$$

3.3.2 Turbulent Flow

As the Reynolds flow number increases above 1200, the system enters a transition from laminar to turbulent flow. Above about 2100, the system is usually turbulent. This results in great mixing, and decreases the settling out of particles. A calm surface layer exists as a function of surface roughness in all turbulent systems enabling fine particles to settle, even from a completely turbulent system. (Larger particles will have a v_s in excess of the upward vertical convection currents.) Davies[5] shows that airborne concentration of particles by number, N_A, at a distance L from the source in a horizontal rectangular duct with turbulent flow and no deposition surfaces other than the floor is

$$N_A = Np \exp\left(-\frac{v_s L}{U h}\right) \tag{3.35}$$

where Np is original concentration in the duct and h is duct height. If specific size particles are generated at a rate of n particles/sec in a duct of width y, initial concentration can be given:

$$Np = \frac{n}{Uhy} \tag{3.36}$$

Fractional deposition efficiency for any given size particle referenced from the initial concentration is

$$\epsilon_i = 1 - \exp\left(-\frac{v_s L}{U h}\right) \tag{3.37}$$

The actual rate of deposition, D_i, at this point for a specific size particle based on initial concentration only is then

$$D_i = \frac{v_s n}{Uhy} \exp\left(-\frac{v_s L}{U h}\right) \tag{3.38}$$

Maximum deposition rate at any given distance occurs when $v_s L = h\bar{U}$.

Figure 3.2 shows how deposition of fine particles in a duct occurs for various gas velocities and particle size expressed as v_s. For particles about 1 μm in air ($v_s \cong 10^{-2}$ cm/sec), it requires a duct length 1000 times the height before more than 5% of these particles are lost. Settling effects increase rapidly because about 36% of 3 μm particles in air ($v_s \cong 10^{-1}$ cm/sec) would drop out in the same distance at 5 mph gas velocities. Deposition by gravitational fallout is inversely proportional to gas velocity.

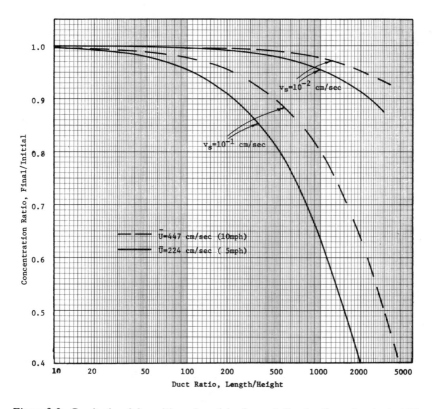

Figure 3.2 Gravitational deposition of particles from air flowing through a duct at SC.

3.4 DIFFUSIVE DEPOSITION

Diffusion becomes more significant as particle size decreases, as noted in Section 2.4 and Table 2.1. Particles so small as to have negligible sedimentation as discussed in the preceding section can have a significant removal rate because of diffusion. If particles in a stationary medium have a residence time in a gravitational field of $t \ll 4D_{PM}/v_s^2$, diffusive deposition would dominate, and if t is much greater, deposition would be by gravitational force.

3.4.1 In a Stationary Medium

Particles reaching a surface are removed by adsorption. If the surface removes every particle striking it, the concentration in the gas phase at the surface is zero and increases with distance from the surface. Not all surfaces are complete adsorbers, but this will be neglected at this time. The original bulk gas concentration of particles is N_P, and the distance from the surface to this location is considered the diffusion boundary layer through which the particles must travel. The thickness of this diffusion layer is δ_d. The ideal rate of deposition for particles (D_i) in a stationary gas is

$$D_i = \frac{D_{PM} \, N_P \, A}{\delta_d} \tag{3.39}$$

where A is the surface area. Deposition velocity due to diffusion is

$$v_p = \frac{D_{PM} \, A}{\delta_d} \tag{3.40}$$

The concentration gradient of particles near a diffusing surface should only be about one-half that for a permeable surface, because particles have an equal chance of returning through a permeable surface.

A comparison of gravitational and diffusional deposition of spherical particles of unit density from still air at SC is given in Table 3.1 for various boundary layer thicknesses.

Table 3.1 Comparison of Terminal Deposition Velocities in cm/sec of Spherical Particles in Still Air at Standard Conditions

Deposition Mechanism	Particle Diameter (μm)					
	1.6	1.0·	0.50	0.10	0.05	0.01
Gravitational deposition ($\rho_p = 1.0$)	8.38×10^{-3}	3.45×10^{-3}	9.89×10^{-4}	8.51×10^{-5}	3.69×10^{-5}	6.61×10^{-6}
Diffusional deposition through boundary layer of thickness:						
1 μm	0.0016	0.0027	0.0062	0.0671	0.233	5.200
10 μm	0.0002	0.0003·	0.0006	0.0067	0.023	0.520
100 μm	—	—	—	0.0070	0.002	0.052
1000 μm	—	—	—	—	—	0.005

Fine particles are often contained in spherical or cylindrical vessels for storage or for observing their properties. While they are contained, the particles can be lost by diffusive deposition as well as by other processes. Assuming gravitational, electrical and other forces are negligible and that the concentration of particles is small enough so that Brownian coagulation is negligible, Pich[6] presents equations showing that diffusive deposition is a function of Fourier number, Fo, when Fo is small numerically, for particles about 0.002-0.2 μm in size. These conditions even apply for containers as small as the alveoli of the lungs, which have an approximate radius of 0.015 cm.   Particles stored in containers for a long time have very small values of Fourier number (*i.e.*, Fo $\ll 1$).

Fourier number is dimensionless and defined for this purpose as

$$Fo = \frac{D_{PM}\, t}{R^2} \tag{3.41}$$

where R is the radius of the cylinder or sphere and t is the time period over which the particle concentration change is being measured. Pich shows that all particles within the mean free displacement distance $\overline{\Delta x_d}$ [see Equation (2.35)], from a surface will be deposited on the surface within some time, t. From this, the ratios of final to initial particle concentration by number for highly disperse particles stored at low Fo are given by Pich, as modified by Slinn,[7] for infinitely long cylinders:

$$\frac{N_A}{N_P} = 1 - \frac{4\,Fo^{1/2}}{\sqrt{\pi}} + Fo + \frac{11}{24}\frac{Fo^{3/2}}{\sqrt{\pi}} + \ldots \tag{3.42}$$

and for cylinders of finite length, L:

$$\frac{N_A}{N_P} = 1 - \frac{4}{\sqrt{\pi}}(1 + \frac{R}{L})Fo^{1/2} + \frac{4}{\pi}(1 + 4\frac{R}{L})Fo - \frac{16}{\pi^{3/2}}\frac{R}{L}Fo^{3/2} \tag{3.43}$$

and for spheres:

$$\frac{N_A}{N_P} = 1 - \frac{6Fo^{1/2}}{\sqrt{\pi}} + 3Fo + \ldots \tag{3.44}$$

Total number of particles deposited in these systems over the time considered becomes equal to container volume times initial number of particles times $(1 - N_A/N_P)$. Deposition of particles on the external surface of shapes can be obtained using the same procedure except that the sign is reversed on the third term on the right in Equations (3.42) through (3.44). For example, total deposition (n_T) of particles in a stationary medium on the outside of a long cylinder is

$$n_T = \pi R^2\, N_P\left[4\left(\frac{Fo}{\pi}\right)^{1/2} + Fo - \frac{11}{24}\left(\frac{Fo^3}{\pi}\right)^{1/2}\right] \tag{3.45}$$

The deposition of particles using the Fourier solution method given here is in good agreement with exact solution methods. For comparison, Table 3.2 shows values of $(1 - N_A/N_P)$ for deposition of particles on the *outside* of a cylinder (the quantity in brackets in the previous equation) calculated by exact solution[1] and by this procedure.

Table 3.2 Relative Number of Particles per Unit Length on the External Surface of an Infinitely Long Cylinder as a Function of Fourier Number (Fo)

Fo	Exact Solution[1]	Slinn Solution[8]
0.001	0.072	0.072
0.005	0.164	0.165
0.01	0.235	0.235
0.02	0.337	0.338
0.05	0.550	0.552
0.1	0.805	0.806
0.2	1.190	1.186
0.4	1.632	1.762

A characteristic time related to diffusion is half-life, $t_{1/2}$, the time required for half the original particles to deposit by diffusion. This is given[6] as

$$t_{1/2} = 0.0305467 \frac{R^2}{D_{PM}} \tag{3.46}$$

3.4.2 From Laminar Flow

The deposition velocity of particles from a laminar gas flowing inside a long cylindrical tube where end effects are negligible is shown[5] to be

$$v_p = \frac{R\,\bar{U}}{2\,L} (1 - \frac{N_A}{N_P}) \tag{3.47}$$

Concentration ratios of particles relative to initial concentration can be found for a long cylinder with laminar flow by Davies'[8] formula

$$\frac{N_A}{N_P} = 0.819 \exp(-14.6272\,C_3) + 0.0976 \exp(-89.22\,C_3) + 0.01896 \exp(-212\,C_3) \tag{3.48}$$

where $C_3 = D_{PM}\,L/(4R^2\,\bar{U})$.

Equation (3.48) is plotted as Figure 3.3 for several values of C_3. Also given in the figure are values of N_A/N_P predicted by Gormley[9] for diffusive deposition of particles from laminar air onto cylindrical and flat surfaces. For the flat surface, $C_3 = 16\,D_{PM}\,L/h^2\,\bar{U}$ where h is the height of the duct or distance between flat plates.

Velocity of diffusional deposition is a function of diffusion layer thickness as given in Equation (3.40). In laminar flow, this thickness is proportional to $(D_{PM})^{1/3}(x/\bar{U})^{1/2}(\mu_g/\rho_g)^{1/6}$ where x is the distance along the surface downwind from entrance. Davies[5] estimates this constant for a flat surface and gives

$$v_p = 0.585\,D_{PM}^{1/3}\left(\frac{\bar{U}}{x}\right)^{1/2}\left(\frac{\rho_g}{\mu_g}\right)^{1/6} \tag{3.49}$$

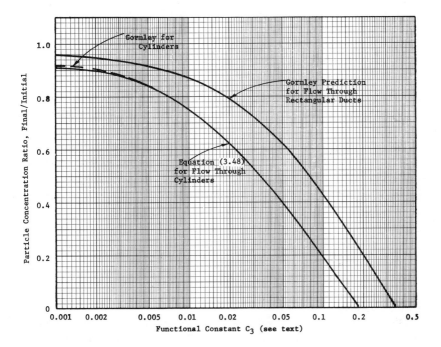

Figure 3.3 Diffusive deposition of particles from a laminar gas stream.

3.5 CURVILINEAR MOTION

Particles approaching an obstacle in a moving gas stream can either move with the gas streamlines and pass by the obstacle or contact the target and be removed, at least temporarily, from the gas. Potential and viscous flow are two basic types of flow used to describe the movement of a fluid around an obstacle. Potential flow assumes the fluid is ideal (*i.e.*, density is constant and viscosity is zero) and that flow is irrotational. Figure 3.4(a) depicts potential flow of a fluid around an infinitely long stationary cylinder. Note that because there is no viscosity, the fluid does not stick to the surface

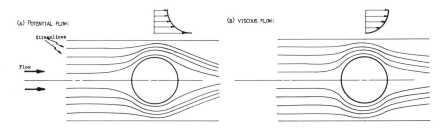

(A) POTENTIAL FLOW:

(B) VISCOUS FLOW:

Streamlines

Flow

Figure 3.4 Streamlines for fluid flow around a long cylinder.

of the cylinder. Any streamline could be replaced by a solid surface with no effects to the flow according to this theory. The velocity profile, shown above the center of the cylinder, shows greatest velocity near the surface.

Viscous flow past a stationary cylinder, as shown in Figure 3.4(b), considers that viscous drag of the fluid at the surface results in a stagnant surface film of fluid. This drag effect is propagated through the fluid resulting in an increasing velocity profile, as shown. For these reasons, viscous flow assumptions are more accurate near the surfaces of objects where viscous effects are important, and potential flow is more correct a short distance removed from the surface.

Actual streamlines do not exist as shown because effects from the presence of the obstacle are propagated both up and downstream causing irregular eddies to exist both before and behind the obstacle. These effects are neglected here.

Collection of particles on the obstacles can result because of impacttion, interception or Brownian diffusion. Particles, because of their greater mass, have more inertia than gas. This causes even fine particles to deviate from the gas streamlines where they may strike the obstacle as shown by particle #1 in Figure 3.5. Particles such as #2 in this figure are removed by interception. These particles do not impact on the target, but are large

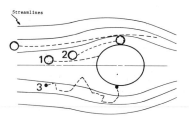

Streamlines

Figure 3.5 Possible movement of particles near a target in a moving gas stream.

enough to contact the target as they pass by and to stick to it. Small particles such as #3 are in random diffusive motion and can be captured by a target when they strike it. This is an important collection mechanism and occurs when particles are removed by filtering the gas using filters with pores larger than the particle. All of these mechanisms can occur whether the gas flow is laminar or turbulent. Detailed treatment of the impaction mechanism is given in Chapter 5.

In curvilinear motion studies, gravitational force usually can be neglected in the presence of the other external forces. Note that if the medium is not at rest, both particle and medium velocities must be considered. This means that from Equation (3.3) we obtain, instead of Equation (3.4), the following expression for particles following Stokes' Law:

$$m_p \frac{dv_p}{dt} = F_E - 3\pi \mu_g d(v_p - \bar{U}) \tag{3.50}$$

For the smaller Cunningham particles, the last expression on the right must be divided by C and for larger particles it is replaced with the Newton Equation as in Equation (3.10).

3.5.1 Oscillating Force Considerations

A pulsing external force can result in oscillation of the particles in the force field. In this case, the external force applied, F_E, is some function of the initial external force, F_{EO}. This can be equated using the appropriate expression, e.g., sinusoidal, rectangular, trapezoidal, triangular, sawtooth, etc. Oscillating electrical fields applied to particles are usually sinusoidal in nature, and F_E can be described (not including F_G) as

$$F_E = F_{EO} \sin (2\pi ft - \phi) \tag{3.51}$$

where f is the frequency of oscillation of the force field and ϕ is the phase shift angle that expresses the lag in oscillation of the particle behind the force application. The lag increases as particle size increases and as force oscillation frequency increases. Abbott[10] showed experimentally that for larger particles

$$\phi \cong \tan^{-1} \tau f \tag{3.52}$$

This is used when particles are > 1 μm in diameter. Smaller particles have increased Brownian motion effects.

When motion due to both gravitational and electrical fields is observed, certain particle properties can be determined. Assume the electrical field is applied horizontally only. The vertical velocity of the particle (v_z), as shown in Figure 3.6, equals terminal settling velocity, v_s or τg. From Equations (3.2) and (2.36) for Stokes particles, this is also

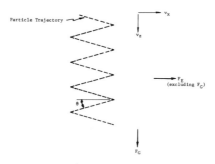

Figure 3.6 Spherical particle falling in a reversing electrical field.

$$v_z = v_s = \tau g = m_p g B \qquad (3.53)$$

and the vertical displacement is

$$z = \tau g t = m_p g B t \qquad (3.54)$$

Hidy[11] shows that the trajectory of the particle produces a horizontal displacement of

$$x = - \frac{F_{EO} \, B \, \cos\left[\left(\dfrac{2\pi f z}{\tau g}\right) - \phi\right]}{2\pi f \left[1 + (2\pi f \tau)^2\right]^{1/2}} \qquad (3.55)$$

The horizontal velocity of the particle is

$$v_x = BEq \qquad (3.56)$$

where E is field strength (potential difference divided by distance between plates) and q is charge on the particle (see Equation 4.11).

Angle of inclusion, θ (see Figure 3.6), is equal to

$$\theta = \tan^{-1} \frac{v_z}{v_x} \qquad (3.57)$$

Photographic measurement of θ and calculation of v_x assuming initially a unit elementary particle charge (a charge of one electron) makes it possible to obtain v_z and hence, a value of particle diameter. Correlation with measurements of x ultimately gives a true particle size and charge, from which exact values of other properties can be calculated.

In a reverse situation, the medium can pulsate and exert an oscillatory force on particles. Small particles, as before, deviate because of their Brownian motion, whereas large particles deviate because of their inertia. Based on a distribution of turbulent energy as a function of Lagrangian frequency, Fuchs[1] notes that about 70% of 0.2 mm particles would be entrained by the medium but only about 2% of 2 mm particles would be entrained.

Turbulent gas flow results in pulsating forces being applied to particles in the medium. For $Re_f > 10,000$, Sherwood and Woertz[12] produced an empirical equation to account for turbulent or eddy diffusivity, D_t, of particles in a rectangular duct:

$$D_t = 0.044 \frac{\mu_g}{\rho_g} Re_f^{0.15} \qquad (3.58)$$

A comparison of v_s/D_t would show whether particles of a given size would settle and stratify in the duct under specific conditions. A ratio near 1.0 would imply no settling and uniform distribution.

3.5.2 Effects of Interception

The impaction collection mechanism implies that a particle is captured on an obstacle because it deviates from the streamline and impacts on the target. If it does not deviate, it passes around the obstacle with the gas and escapes. However, a particle of finite size can move with a streamline yet be collected on the obstacle if the particle radius is as large or larger than the distance the closest streamline is displaced. This is shown in Figure 3.7 where δ is this streamline displacement distance.

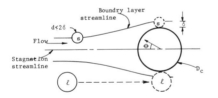
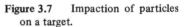

Figure 3.7   Impaction of particles on a target.

Close to the obstacle surface and at low Re_f, viscous flow assumptions are more accurate. Farther from the surface and at large Re_f, potential flow assumptions are better. It is assumed that velocity of the medium is zero within the boundary formed by the closest streamline and that the boundary layer (streamline) remains laminar up to Re_f of 10^5, at which time it breaks away from the obstacle surface. The stagnation point is defined when θ in Figure 3.7 is $0°$ and the stagnation streamline is the obstacle centerline. Interception occurs over the surface from θ of $60-80°$.

Interception results in an increase in particle collection over that predicted by impaction alone. This efficiency increase is a function of the ratio of particle to collector diameters, d/D_c. Estimations of this increase in fractional collection efficiency, $\Delta\epsilon_i$, for various size particles due to impaction on cylinders and spheres have been made[1] and are given below for various cases. Where particles are large, Stokes' number ($St = 2x_s/D_c$) approaches infinity. Under these conditions the particle moves in a straight line as shown by particle "ℓ" in Figure 3.7. For this case, $\Delta\epsilon_i$ becomes

$$\Delta\epsilon_i = d/D_c \qquad\qquad \text{for cylinders,} \qquad\qquad (3.59)$$

and

$$\Delta\epsilon_i \cong \frac{2d}{D_c} \qquad\qquad \text{for spheres.} \qquad\qquad (3.60)$$

Small particles with $d/D_c \ll 1$ follow the streamline as shown by "s" in Figure 3.7. For these cases, $\Delta\epsilon_i$ becomes

$$\Delta\epsilon_i \cong \frac{2d}{D_c} \qquad\qquad \begin{array}{l}\text{for cylinders assuming} \\ \text{potential flow,}\end{array} \qquad (3.61)$$

$$\Delta\epsilon_i \cong \frac{(d/D_c)^2}{2.002 - \ln Re_f} \qquad \begin{array}{l}\text{for cylinders assuming} \\ \text{viscous flow } (Re_f < 1),\end{array} \qquad (3.62)$$

$$\Delta\epsilon_i \cong \frac{3d}{D_c} \qquad\qquad \begin{array}{l}\text{for spheres assuming} \\ \text{potential flow,}\end{array} \qquad (3.63)$$

and is

$$\Delta\epsilon_i \cong \frac{3}{2}\left(\frac{d}{D_c}\right)^2 \qquad \begin{array}{l}\text{for spheres assuming} \\ \text{viscous flow } (Re_f < 1).\end{array} \qquad (3.64)$$

Overall efficiency estimates require iterative calculations for various size particles using the appropriate equations for the various size ranges, Re_f and type of collector. Fine particles will fall in the $d/D_c \ll 1$ group.

3.6 SCALE-UP OF SYSTEMS

It is important to consider scale-up factors so that pilot plant systems can be expanded for full-size use. Actual space available, desired system capacity and material of construction limitations coupled with such factors as hold-up time, volumetric flow rate, pressure and temperature, may dictate ultimate size extremes. Even so, other factors need to be considered. One is the general consideration that a custom-designed, properly made facility should work best for each specific situation. This is especially important for large facilities. Scale-up of individual components within a system requires incorporation of the following generalizations.

A direct scale-up design for components handling flowing gas streams would include using the same gaseous medium and particulates and operating at the same temperature, humidity, pressure, etc. as the test system. The second requirement is that both systems be geometrically similar, and the third is that there be similarity of motion. Similarity of motion can be obtained by making certain each unit has equal values of the dimensionless Reynolds flow number, Re_f, Stokes number, St, and Froude number, Fr. Re_f is the ratio of inertial to viscous forces. Fr is the ratio of inertial to gravitational forces and need only be considered if gravitational force is important. St is proportional to the ratio of particle stopping distance to diameter of collector and, as such, is a ratio of inertial force to characteristic dimension of collector. The stopping distance must be corrected for deviation from Stokes' Law as described in Section 3.2.2.

These dimensionless numbers are given in Section 1.6 and are summarized here for emphasis:

$$Re_f = \frac{D \, v_g \, \rho_g}{\mu_g} \qquad (1.9)$$

$$St = \frac{2x_s}{D} = \frac{2\tau \, v_g}{D} = \frac{d^2 \, \rho_p \, v_g}{9 \, \mu_g \, D} \qquad (1.19)$$

$$Fr = \frac{v_g^2}{Dg} \qquad (1.20)$$

In Equation (1.9), D is diameter, or equivalent, of the system, whereas in Equation (1.19) D is the collector dimension, which may or may not be the system diameter. If one considers only inertial forces, with all other physical properties of the substances being constant and dimensions of collector equal to system diameter, then the ratio of $(St/Re_f)^{1/2}$ would be a constant. Inertial collection only in pilot and full-size systems would both have collection efficiencies $\propto d/D$. Note that under these restrictions, the larger system with larger D would have a lower inertial efficiency than the small system, and a design variation would be needed to increase this efficiency.

REFERENCES

1. Fuchs, N. A. *Mechanics of Aerosols* (London: Pergamon Press, 1964).
2. Bird, R. B., W. E. Stewart and E. N. Lightfoot. *Transport Phenomena* (New York: John Wiley & Sons, Inc., 1965).
3. Hesketh, H. E. *Understanding and Controlling Air Pollution*, 2nd ed. (Ann Arbor, Mich.: Ann Arbor Science Publishers, Inc., 1974).

4. Thomas, J. W. "Gravity Settling of Particles in a Horizontal Tube," *J. Air Poll. Control Assoc.* 8(1):32-34 (1958).
5. Davies, C. N. *Aerosol Science* (New York: Academic Press, 1966).
6. Pich, J. "Theory of Diffusive Deposition of Particles in a Sphere and in a Cylinder at Small Fourier Numbers," *Atmos. Environ.* 10(2):131 (1976).
7. Slinn, W. G. N. "Theory of Diffusive Deposition of Particles in a Sphere and in a Cylinder at Small Fourier Numbers," *Atmos. Environ.* 10(9): 789 (1976).
8. Davies, C. N. *Proc. Royal Soc.* B133:298 (1946).
9. Gormley, P. and M. Kennedy. *Proc. Royal Irish Acad.* 52A:162 (1949).
10. Abbott, R. *Phys. Rev.* 12:381 (1918).
11. Hidy, G. M. and J. R. Brock. *The Dynamics of Aerocolloidal Systems* (London: Pergamon Press, 1970).
12. Sherwood, T. and B. Woertz. *Ind. Eng. Chem.* 31:1034 (1939).

CHAPTER 4

EFFECTS OF OTHER FORCES

Some effects of various forces have been presented in the preceding discussions to assist in developing basic techniques to describe particle behavior and motion. In addition, forces important in the control of fine particulate pollutants are included in Chapter 5 (Particle Collection) with the respective devices that make use of these forces. This chapter presents particle properties relevant to the various forces and how the particles are influenced because of these properties. Not all forces can be covered in this discussion because not enough information is available. One example is the mysterious "pyramid force" that is claimed to exist and produce unusual effects. Some of the forces discussed here are "mysterious" enough in their own way to produce pyramid results if we evaluate them fully.

4.1 INERTIAL FORCE

4.1.1 General

Impaction of a particle in a moving gas stream on an obstacle can occur by inertia, interception or Brownian diffusion, which is noted in Section 3.5. Inertial impaction is the mechanism most frequently used to remove particulate matter. Corrections to account for interception as described in Section 3.5.2 and for diffusion, as in Chapter 5, must be considered and used if necessary.

Particles have a much greater mass, and therefore much greater inertia when in motion, than the surrounding gas. Consider for the moment that each particle is an infinitely small point source so interception need not be considered. The gas has a low density and passes around the obstacle collector with the diameter D_c in streamlines as shown in Figure 4.1. The heavier particles resist the change in momentum and cross the gas streamlines

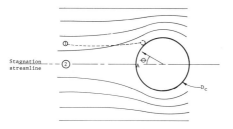

Figure 4.1   Inertial impaction of particles from a moving gas stream onto a target.

striking the obstacle as shown by particle "1." This collecting target can be stationary or in motion, but there must be a velocity difference between the particle and the target. If the particle is moving slower and the obstacle overtakes it, this is called scavenging and is treated in Section 4.1.4.

Particles that impact on a target may stick to it or may escape by bouncing or being blown off. Greater velocity differences are required between particle and obstacle to cause smaller particles to impact, but this also increases bounce and blow-off. Both adhesive and cohesive forces are important in determining whether the particle remains attached. The initial assumption is that all particles adhere when they strike the target.

The stagnation streamline is the centerline of the obstacle with respect to the flow of the gaseous medium (Figure 4.1). The stagnation point occurs on the surface of the obstacle at A; *i.e.,* where $\theta = 0°$. At an infinite distance from the obstacle, a particle on the stagnation streamline is assumed to move at the velocity of the gas and is not influenced by the presence of the obstacle. As it approaches the obstacle, the gas must pass on either side of the target leaving the particle propelled toward the obstacle by inertia only. This particle will only reach the target if it has adequate momentum to cross the streamlines. This can be determined by the stopping distance techniques given in Section 3.2.2. Inertial collection of this particle is thus related to stopping distance and collector dimension, which is Stokes number as given by Equation (1.19). The critical Stokes number, St_{cr}, exists for particles approaching on the stagnation streamline. For all other particles, collection occurs *only* when $St > St_{cr}$ because all other particles have a tangential velocity component as well as a normal component.

Hidy[1] presents values of St_{cr} based on potential flow theory at high values of Re_f as determined for various shapes. Particles with St less than the St_{cr} noted should not impact on the obstacle, because of inertial force. These theoretical values compared with approximate experimental values of St_{cr} are given in Table 4.1. These comparisons show the many problems that still exist in establishing the proper theoretical approach.

Inertial impaction collection efficiency data are frequently reported in terms of inertial impaction parameter, K_I. The reader is cautioned in using

Table 4.1 Comparison of Experimental and Theoretical Values of
Minimum Critical Stokes Number Required for Collection
of Particles on Various Surfaces

| | St_{cr} | |
	Experimental Effective	Theoretical
Jet stream perpendicular to plane (for large ℓ/w; see Figure 4.2)	0.16	$2/\pi$
Stream approaching circular disc	0.2	$\pi/16$
Stream approaching sphere	0.1	$1/12$
Stream approaching cylinder (with projected area A_p = major axis/minor axis)	0.3	$\dfrac{1}{4(1 + A_p)}$

this term because various forms are in common use. In this book impaction parameter is the same as Stokes number, as defined. Typical values in literature are the equivalent of ½ St and $\sqrt{½\,St}$. Check individual definitions before using. The Stokes number or impaction parameter used here as expressed in the most common form for spherical fine particles that obey Cunningham's correction is

$$K_I = St = \frac{d^2\,\rho_p\,v\,C}{9\,\mu_g\,D_c} \tag{4.1}$$

where particle and gas velocities are assumed equal. Other shape and size particles must be corrected for accordingly.

4.1.2 Impingement

Commercially available impingers utilize inertial impaction to remove heavy concentrations of particulates. Two arrangements are shown in Figure 4.2. These are the Greenburg-Smith type and can be operated dry or with liquids. The plated impinger insert is best for particle removal and operates at about 30 lpm gas flow. The nozzle width (w) is 2.3 mm giving an exit jet velocity of about 114 m/sec at 1.0 cfm (28.32 lpm) flow rate. The distance from the jet to the strike plate (ℓ) is 4 mm. The straight-end impinger has values of w = 10 mm and ℓ = 8 mm for a jet velocity of 9.4 m/sec. Normally this one is operated under about 75 ml of liquid. Both impingers are in 57 mm cylinders 425 mm tall. A commercial midget impinger (not shown) is similar to Figure 4.2(b) except that the strike plate is the bottom of the

(A)   STRAIGHT END
      BUBBLER INSERT

(B)   PLATED IMPINGER
      INSERT

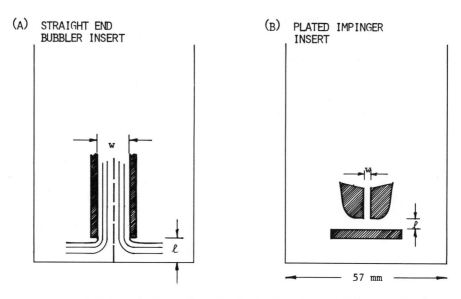

Figure 4.2   Commercially available inertial impingers for collecting particles
from a jet on a flat surface.

26 mm x 175 mm cylinder.   Values of w and ℓ are 0.992 mm and 5 mm,
respectively, giving a jet velocity of 61 m/sec at 0.1 cfm (2.83 lpm).

Gas streamlines are depicted in Figure 4.2(a) with the dashed center line
representing the stagnation streamline.   As before, efficiency for any particle
size is proportional to stopping distance divided by characteristic dimension;
i.e., proportional to Stokes number.   In this case, the characteristic dimension
is represented by w. Ranz and Wong[2] data best fit experimental data for these
impinger devices and are plotted as the solid line in Figure 4.3.   The dashed
line represents a linear approximation of impaction collection efficiency on a
flat surface:

$$\epsilon_i = 1.15 \text{ St} - 0.24 \qquad (4.2)$$

where ϵ_i is individual fractional collection efficiency.   Ranz and Wong experimental collection efficiency data for spheres, and Lundgren[3] experimental data for cylinders are also given in Figure 4.3.   The Ranz and Wong
sphere data and Lundgren cylinder data for St > 6 are higher than those
predicted by potential flow theories.

The approximate inertial impaction efficiency on spheres is then

$$\epsilon_i = 0.203 \ (5.04)^{\text{St}} \qquad (4.3)$$

and on cylinders is

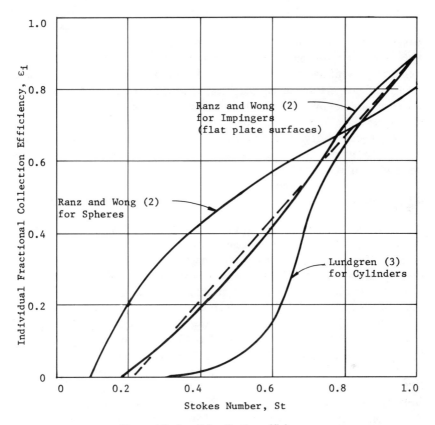

Figure 4.3 Inertial collection efficiency.

$$\epsilon_i = 3.2 \times 10^{-4} \exp (10.5 \ St) \text{ when } St \leqslant 0.7 \qquad (4.4)$$

and

$$\epsilon_i = 0.094 \ (9.72)^{St} \text{ when } 0.7 < St < 1.04 \qquad (4.5)$$

Equation (4.5) approximates the theoretical data for collection efficiency on cylinders for higher values of St.

One reason why experimental data vary from theoretical could be the concentrating effect of particles in the centers of converging streams. Inertial impaction devices usually function by forcing the gaseous medium into a venturi-type reducing section to obtain a high-velocity jet. If the sides of the jet-forming nozzle were long and parallel, the particle distribution would be uniform, as discussed in Section 3.3.2. As the sides converge, smaller particles move toward the axis at an angle greater than the converging angle. Large particles are not displaced and move toward the axis only as necessary geometrically. Other reasons why experimental and theoretical data do not agree include problems such as measuring particle concentration in the medium, preventing the particles from leaving the surface once impaction occurs and coagulation of the particles in the gas stream. Moving targets can scavenge particles by collecting them on the "back" or downstream side when the targets move faster than the particles.

Impingement devices can be partially filled with liquid, which enhances particle collection; this is discussed in the following section on scrubbers under collection by bubbles.

4.1.3 Scrubbing

Collection of particles on droplets of liquid is of extreme importance in scrubbing devices. Goldshmid and Calvert[4] showed that shape and movement of the collecting droplets have a negligible effect on impaction collection efficiency. This study did show that at low St (0.1-0.55) collection on a droplet was influenced by the wettability of the particle. Hydrophobic substances experienced a lower collection efficiency than hydrophilic particles. Also, at low St submicron particles at high velocities are frequently captured on the back side of the droplets. These behaviors can be explained by consideration of phoretic forces (later in this chapter).

In scrubbers where the droplet targets move, complete collection efficiency is assumed when the droplet sweeps clean an area of the gas medium equal to the droplet projected area. Proper atomization and distribution of the droplets are required to try and fill the entire cross-sectional area so all the gas is scrubbed. The use of deflected gas streamlines makes it possible to define a limiting streamline based on actual impaction collection efficiency. This limiting streamline is the farthest distance from the stagnation streamline a particle can be and still be captured by impaction. Anything farther away could only be captured by interception or diffusion. This distance varies for each size particle, and in symbols, is related to efficiency:

$$\epsilon_i = \left(\frac{2Y_0}{D_c} \right)^2 \quad \text{for spheres} \tag{4.6a}$$

and

$$\epsilon_i = \frac{2Y_0}{D_c} \quad \text{for cylinders} \tag{4.6b}$$

where Y_0 is the distance from stagnation to limiting streamline and D_c is collector diameter. This definition for fractional collection efficiency means that values over 1.0 indicate other forces (*e.g.*, electrical) are causing particles to move toward the droplets.

Not all scrubbers operate with atomized droplets as collectors. Sieve plates, for example, consist of perforated trays upon which downward-moving liquid is suspended by the force of the rising gas. The gas rising from the holes in the plate bubbles through the liquid. The particles are temporarily trapped inside the bubbles and some are captured. Impingers and bubblers act similarly except that gas flow is usually downward in impinger devices and there may be no liquid flow.

Particle collection can occur during bubble formation or during bubble rise. It has been determined[5] that most collection occurs during bubble formation where the collecting surface is the continuously changing semi-spherical dome formed by the inside bubble wall. Collection efficiencies for particles in the Continuum and Slip Flow Regimes can be estimated using[5]

$$\epsilon_i = 1 - \exp(-a \, St) \qquad (4.7)$$

where the value of St can be obtained from Equation (4.1) using, for example, the plate perforation diameter for D_c and the gas velocity through the perforation for v_g. The constant "a" is related to foam density, which is considered to be the ratio of quiet liquid level to bubbling liquid level. For hydrophilic particles, "a" equals 40 times the square of foam density. It must be obtained experimentally for a hydrophobic particle. The smaller particles are also captured by diffusive force and are discussed in Section 4.5.

4.1.4 Scavenging

Removal of particles from gas by droplets falling through the medium is called scavenging. It is an impaction process in the sense that the particles removed do impact on the droplets. Inertial impaction or aerodynamic capture exists, but in addition, other processes such as particle Brownian diffusion, particle acting as a condensation nuclei, supersaturation of the gas, phoretic forces, ultrasonics and electrical charges can help. Inertial impaction, however, does dominate for particles larger than 1 μm. The other factors listed become more significant for smaller submicron particles and are covered in the sections that follow.

Scavenging is an important process for both atmospheric cleaning and particulate collection equipment. Scavenging also exists in the discharge section of high-energy scrubbing devices such as venturis, where the large accelerated droplets in the diverger decelerate more slowly than the gas and small particles. In the dewatering section required of high-energy scrubbers, scavenging occurs when the droplets are forced to pass through the gas and remove particles. Countercurrent, low-energy scrubbers such as spray towers utilize falling droplets to clean the particulates from the rising gas.

In this discussion on scavenging it is assumed, as usual, that both particles and droplets are spherical and that once contacted they do not separate. Collection by droplets falling at high Re_f is best described by potential flow theory, but this neglects the turbulent eddies formed at the back side of the droplet. Until better data are available, collection efficiency of larger droplets should be estimated using the inertial impaction equations. Always consider the larger (particle or droplet) as the collector.

Very small droplets would have a low Re_f and fall into the viscous flow theory. Therefore, both submicron particles and small droplets result in Brownian diffusion capture dominating. This is further discussed in Section 4.5. Kerker[6] reports this to be the case when droplets less than about 2.5 mm in diameter are used to scavange submicron silver chloride particles. The inertial impaction mechanism applies for larger drops. For < 2.28 mm droplets in free fall, fractional collection efficiency for the droplet on all particles is

$$\epsilon_i = 1.68 \, Pe^{-2/3} \tag{4.8}$$

where Pe is Peclet number, defined

$$Pe = \frac{v_s \, D_C}{D_{PM}} \tag{4.9}$$

The falling droplet velocity is given as v_s. Note that this efficiency is very low, being about 1×10^{-5}.

For droplets less than about 2.5 mm, scavenging collection efficiency on *larger* particles, according to the data of Adams and Semonin,[7] increases with decreasing droplet size:

$$\epsilon_i = 0.0077 \, \exp(-2.56D_1) \tag{4.10}$$

where D_1 is droplet size in mm.

4.2 ELECTROSTATIC FORCE

4.2.1 Particle Charge

Gases are poor conductors of electricity, and because of this, particles in neutral gaseous mediums can have (a) no charge, (b) positive charge, or (c) negative charge. Apparent charges can be induced on electrically neutral particles by the presence of charged surfaces or charged gaseous ions. This is shown schematically in Figure 4.4. The presence of charged objects can

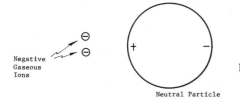

Figure 4.4 Schematic of dipole moment induced on a neutral particle by the presence of charged ions.

even induce an apparent charge of different magnitude on already charged particles.

Particles can receive an electrostatic charge by one or more of three basic procedures: (a) during generation, (b) subsequent contact and release from charged surfaces, and (c) gaseous ion diffusion. There are other possible procedures such as direct charging in the gaseous medium by free electrons arcing across an electrical potential, but these are negligible compared to the three basic ways. Ion diffusion is the dominant technique by which particles are charged.

Particles are charged during generation by

1) Electrolytic charging—electrons are exchanged at the surface of a high dielectric liquid-solid interface followed by separation of the liquid; this can occur during atomization. Water is a poor conductor of electricity and charged atomized droplets can be readily formed.

2) Spray electrification—a charged liquid can be separated, *e.g.*, by atomization, to form charged droplets.

3) Contact separation—separation of contacting, dry, nonmetallic and, therefore, nonconducting surfaces, *e.g.*, by grinding; this is also known as tribo electrification.

Subsequent charging of already-formed particles by contacting and then releasing from a charged surface is called contact charging. This happens when particles deposit on a charged surface and then are knocked or blown free. Electrostatic precipitator rebound and reatomization from a charged surface are examples of this.

Ion diffusion charging occurs when gaseous ions, such as in Figure 4.4, contact the particles. The induced charges retain the ion electrical charge, and in this case, result in a net negative charge on the particle. Gaseous ions can be created by electrical discharges or corona, radioactivity and flame, cosmic or photoionization. Corona charging of particles is important in the operation of electrostatic precipitators while all of the procedures are being used to enhance operation of scrubbers by charging either or both the collecting droplets and the particles.

A corona consists of an electrically active region in the vicinity of a discharge electrode where electrons are stripped from neutral gas molecules. In most industrial electrostatic precipitators the electrode is negative so the ions formed *directly* are positive. These are attracted to the discharge electrode while the free electrons are accelerated by the electric field until they reach a velocity adequate to ionize neutral gas molecules by impact. These ions are negative. In the corona glow region there is an avalanche process that occurs when naturally occurring free electrons are used to produce the initial ions and subsequent electrons. In contrast to the industrial processes, a positive corona is used in inhabited areas (*e.g.*, for air conditioning cleaning) because of the reduced production of ozone. The ionization potential necessary to produce the corona for air and most common gases in the air is from 12.2-15.5 V.

The impaction of many free electrons on a gas molecule is required to form a single negative ion. The average number of collisions required to form single ions[8] is approximately:

Inert gases, N_2 and H_2	∞
CO	1.6×10^8
Air	4.3×10^4
H_2O	4.0×10^4
O_2	8.7×10^3
SO_2	3.5×10^3

The movement of any free electron, ion or particle through a nonconductive gas is related to the electric field strength (E) by Coulomb's Law:

$$F_e = qE = n_p eE \qquad (4.11)$$

where q is the net electrical charge on the object, F_e is the electrical force, n_p is the number of electron charges and e is the charge on one electron, which is 1.603×10^{-19} coulomb. Field strength equals electrical potential divided by plate spacing. The electrostatic force results in a migration velocity (v_e) of

$$v_e = BqE = bE \qquad (4.12)$$

where b is electrical mobility of the charge carrier. The mobility of electrons $\gg$ mobility of ions $\gg$ mobility of charged particles. The electric mobility of some gaseous ions is given in Table 4.2. Note that the inert gases, N_2 and H_2, can ultimately be charged positively and, therefore, have positive ion mobility as noted.

Equation (4.12) is a defining equation for electrical mobility, *i.e.*, $b = v_e/E$. It also shows that mechanical mobility B and electrical mobility are related by

Table 4.2 Electrical Mobility of Single-Charged Gaseous Ions at $0°C$ and 1 atm

Gas	Electric Mobility, b [$m^2/(sec\text{-}V)$]	
	Negative Ion	Positive Ion
He	—	10.4×10^{-4}
Ne	—	4.2
Ar	—	1.6
Kr	—	0.9
Xe	—	0.6
Air (dry)	2.1×10^{-4}	—
Air (very dry)	2.5	1.8
N_2	—	1.8
O_2	2.6	2.2
H_2	—	$12.3(H_3^+)$
Cl_2	0.74	0.74
CCl_4	0.31	0.30
CO	1.14	1.10
CO_2 (dry)	0.98	0.84
H_2O ($100°C$)	0.95	1.1
H_2S	0.56	0.62
NH_3	0.66	0.56
N_2O	0.90	0.82
SO_2	0.41	0.41

$$b = n_p e B \qquad (4.13)$$

It is necessary to properly resolve all units noting that one coulomb equals 10^7 g $cm^2/(sec^2$ volt). The migration velocity, v_e, of a charged spherical particle in quasistationary motion toward the collecting electrode when the viscous drag force of the medium is equal to the applied electrical force can be obtained from Stokes' Law corrected by the Cunningham factor:

$$v_e = \frac{n_p e E \; C}{3\pi \, \mu_g \, d} \qquad (4.14)$$

Under these conditions, b would equal

$$b = \frac{n_p e \; C}{3\pi \, \mu_g \, d} \qquad (4.15)$$

and have the units of $cm^2/(V \; sec)$. Similarly, $B = C/3\pi\mu_g d$ with the units sec/g. The most frequent number of electron charges on a particle is unity; however, this depends on particle size and other factors as discussed below so the best value of n_p should be used. A terminal or maximum migrational velocity occurs for any given charge when the field strength of about 10,000 V/cm is used. This is about the maximum before air breaks down.

The maximum amount of charge acquired by a particle is limited by the electrical breakdown strength of the surrounding medium. Breakdown strength for dry air is about 8 esu/cm^2 (or 1.66×10^{10} electrons/cm^2 as one electron = 4.8×10^{-10} electrostatic units). Billings and Wilder[9] show the estimated maximum charge on various size particles as a summary of 36 references. This is given as Figure 4.5. They also conclude that normal observed charges are about one-tenth the maximum value. Highly charged particles can emit electrons (negative particles) or ions (positive particles), and liquid drops can disintegrate.

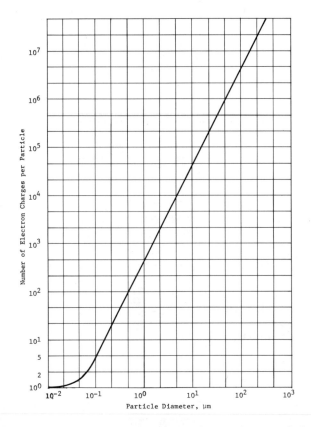

Figure 4.5 Calculated and measured maximum electron charge on single particles in dry air.[9]

Normal particles in the atmosphere are exposed to and in equilibrium with naturally charged (bipolar) ions. These ions are produced by cosmic, radioactive and other natural ionization processes. As a result, the particles become charged by ion diffusion and achieve a charge distribution as estimated by Whitby and Liu[10] (see Table 4.3). Note that for particles > 0.1 μm, the median number of charges expected is one electron.

When particles are placed in an electrical field, two physical mechanisms are mainly responsible for transporting charges to the particles. These are field and diffusional charging. Field-dependent charging consists of the electrical field (coulombic force) providing the corona to generate electrons and ions, which are accelerated by the electrostatic driving force (as described above) to ultimately charge the particles. Diffusion charging consists of the random motion of both the charged ions and the particles to cause collision and charging.

A charge transported to the vicinity of a particle by any of the mechanisms discussed induces a charge on the particle, as in Figure 4.4. This induced charge is called the image charge force and is what holds the charge received during the collision. Once charged, the particle repells like charges and reduces the probability of further collisions and subsequent charging. If a saturation charge is acquired by a particle, the electrical field produced by the charged particle will equal the applied field, and no more ions can be driven onto the particle without an increase in the field strength.

Particle charging rate is related to free ion density in the vicinity of the particle and ion mobility. The charging time constant for field-dependent charging (t_c) is the time required for a particle to reach one-half the saturated value:[8]

$$t_c = \frac{4\,\epsilon_0}{N_0 e b} \tag{4.16}$$

where ϵ_0 is the dielectric constant of vacuum, which is 8.8 x 10^{-12} coulombs^2/(joule m).

Particles larger than about 0.8 μm in size are dominantly charged by field charging. For these particles with a relative dielectric (ϵ) constant greater than one, the saturation value of field charge (q_s) is[8]

$$q_s = 3\pi\left(\frac{\epsilon}{\epsilon+2}\right)\epsilon_0\, d^2\, E \tag{4.17}$$

The charge as a function of time (t) for field charging alone is

$$n_p = \frac{q_s}{e}\left(\frac{t}{t+t_a}\right) \tag{4.18}$$

Table 4.3   Fractional Equilibrium Distribution of Charges on Spherical Particles in a Natural Bipolar Ion Atmosphere[10]

d, μm	Number of Electron Charges on Particle, n_p											Average Charge
	0	1	2	3	4	5	6	7	8	9	10	
0.01	0.993	0.007										0.007
0.015	0.955	0.045										0.045
0.02	0.900	0.100										0.100
0.03	0.763	0.236	0.001									0.238
0.06	0.550	0.430	0.020									0.470
0.1	0.424	0.480	0.090	0.006								0.678
0.3	0.241	0.410	0.232	0.093	0.024	0.005						1.247
1.0	0.133	0.253	0.214	0.162	0.109	0.065	0.035	0.017	0.007	0.003	0.001	2.350

where n_p is the number of electron charges and t_a is charging or acceleration time constant equal to $1/(\pi e\ N_o\ b)$. When ionic mobility (b) of air is used, $t_a \cong 1 \times 10^6/N_o$ in seconds. N_o is ion concentration per cm^3. An example of use of these equations is given below.

The small ($< 0.2\ \mu m$) particles are charged mainly by unipolar ions in time t by diffusion charging alone by the amount[10]

$$n_p = \frac{dkT}{2e^2}\ \ell n \left[1 + \frac{\pi d\ v_i\ N_o\ e^2\ t}{2kt}\right] \qquad (4.19)$$

where k = Boltzmann constant, 1.38×10^{-16} ergs/(molecule $^\circ$K)
$\quad\quad\ N_O$ = initial ion density, ions/cm^3
$\quad\quad\ v_i$ = root mean square ion velocity, cm/sec
$\quad\quad\ t$ = time, sec
$\quad\quad\ T$ = temperature, $^\circ$K

The root mean square ion velocity is given by

$$v_i = \left(\frac{8kt}{\pi m}\right)^{1/2} \qquad (4.20)$$

where m is the mass of the ion.

Charging occurs due to both mechanisms in the intermediate size range, and a minimum particle charge occurs at about 0.3 μm. This is shown by the calculations of Lowe and Lucas[11] as given in Table 4.4. The limiting field charge at t $= \infty$ corresponds to that given in Figure 4.5 except that Figure 4.5 assumes maximum field strength before air breaks down. These calculations were made assuming high particle dielectric constant, a wire and tube-type precipitator at 300 $^\circ$K with $N_o = 5 \times 10^7$ ions/cm^3 and $E = 2000$ V/cm in air. The voltage assumed is 40 kV at 1.3 μamp/cm discharge electrode current.

Table 4.4 Calculated Unit Charges Acquired by Particles[11]

Particle Diameter (μm)	Exposure Time (sec)							
	Field Charging				Diffusion Charging			
	0.01	0.1	1	∞	0.01	0.1	1	10
0.2	0.7	2	2.4	2.5	3	7	11	15
2.0	72	200	244	250	70	110	150	190
20.0	7200	20,000	24,000	25,000	1100	1500	1900	2300

An example follows to show how saturation charge on a particle can be calculated using Equation (4.17). Assume that a limiting field strength of 10,000 V/cm is used; the particle size is 2 μm, the particle relative dielectric is 4 and the available charging time is "long." Use of the conversions 10^7 g cm^2/(sec^2 V) per coulomb, 1.602 x 10^{-19} electrons/coulomb and 100 cm/m gives the factor 6.24 x 10^{16}, which resolves units of ϵ_0 in coulombs^2/(joule m), d in cm, and E in volts/cm to give the saturation charge by field charging in electrons in terms of number of electrons:

$$n_p = (3)(\pi) \left(\frac{4}{4+2} \right) (8.85 \times 10^{-12})(2 \times 10^{-4})^2 (10,000 \text{ V/cm})(6.24 \times 10^{16})$$

$$= 1380 \text{ electrons}$$

which agrees closely with Figure 4.5. Saturation charge q_s then equals $(1380)(1.602 \times 10^{-19}) = 2.21 \times 10^{-16}$ coulombs.

Bipolar charged particles < 10 μm tend to clump together. This can be good if the particles are to be removed from a gas, but it can be bad for sieving, drying and other processes where the particles must ultimately be removed from the device. In contrast, unipolar charged particles maintain their charge for long periods of time. Electrostatic repulsion keeps them from coagulating so they are considered to remain charged until they precipitate on the surrounding walls. Whitby[12] shows that the $t_{1/2}$ of decaying particles in a spherical system of radius R can be found from

$$t_{1/2} = 0.691 \frac{3\pi \mu_g d (R)^{3/2}}{qC} \left(\frac{b}{6Q_i} \right)^{1/2} \tag{4.21}$$

Q_i is the ion generation rate and $t_{1/2}$ is the time for particle concentration to reach one-half the initial value.

Estimates of electrical charge on particles can be made by measurement techniques such as the one described in Section 3.5.1. A summary of the state of this art as of 1970 was made by Lapple.[13] Many of these procedures require extreme care and great patience on the part of the observer.

4.2.2 Deposition

The previous equation shows that half life of unipolar particles depositing on surfaces is a function of particle size and electrical mobility. Mobility is the more significant variable as this decreases about 10 times faster than size increases. Therefore, smaller particles have a much shorter half life. Experiments in a 2000-ft^3 room[12] showed that particles about 0.05 μm in size have a half-life of 5.5 minutes while particles 3.6 μm have a half life of nearly one hour. This shows how small charged particles are significantly influenced by electrostatic forces but large particles are not.

The effect of charge on movement of particles can also be shown by a force balance as in Section 3.2.1. However, this time the external force is the electrical field, and gravitational force can be assumed to be negligible. The resultant nonsteady-state velocity of a particle starting from rest is equivalent to Equation (3.8) where the steady-state velocity of a charged particle is given by Equation (4.14):

$$v_e = \frac{n_p \, eE \, C}{3\pi \, \mu_g \, d} \, (1 - \exp(-t/\tau)) \tag{4.22}$$

where τ is the relaxation time constant as given by Equation (3.2). Very small particles with their extremely low relaxation time (see Table 2.1) reach terminal velocity quickly so the value of $\exp(-t/\tau)$ can be neglected and collection efficiency is high because the particles have much time to move to the collectors. Large particles have a high relaxation time so the effective acceleration period is short compared with overall residence time. This could result in a collection efficiency decrease if the system is not designed properly. If a mechanical precleaner or other precautions are taken to remove the large particles before the precipitator, Equation (4.22) for practical systems becomes Equation (4.14).

ESP

Particle migration velocity for electrostatic precipitator (ESP) systems, where the very large particles have been eliminated before entering the precipitator, can be expressed in terms of charging time by combinations of the charging and velocity equations. Typical residence times for commercial ESPs range from 2-5 seconds. For particles > 0.8 μm not removed by precleaning, field-dependent charging dominates and migration velocity is

$$v_e = \left(\frac{\epsilon}{\epsilon + 2}\right)\left(\frac{\epsilon_0 \, E_0 \, E_p \, d}{\mu_g}\right)\left(\frac{1}{1 + \tau/t}\right) \tag{4.23}$$

Particles with high relative dielectric constant makes $\epsilon/(\epsilon + 2) \to 1$ and for long times when saturation charge is attained $\tau/t \to 0$ and Equation (4.23) gives the steady-state velocity for field charging:

$$v_e = \frac{\epsilon_0 \, E_0 \, E_p \, d}{\mu_g} \tag{4.24}$$

where E_0 and E_p are charging and precipitating field strengths respectively. In single-stage units, E_0 equals E_p. Units and appropriate resolution are explained in the example below. The maximum field strength at which air breaks down is about 10,000 V/cm.

Diffusional charging dominates for particles < 0.2 μm, and migration velocity for this becomes

$$v_e = \frac{E_p}{6\pi\,\mu_g} \frac{kT}{e}\, \ell n \left(\frac{2 + \pi d\, U_O\, N_O\, e^2 t}{2kT} \right) \qquad (4.25a)$$

Saturation charges are not approached by diffusional charging and most of this occurs within 20-50 msec in commercial units. Equation (4.25a) is difficult to use so another expression is given using Cochet's[14] modification. This can be used to obtain migration velocities due to diffusional charging and also obtains a correction to account for both types of charging in the region of particle size from 0.2-0.8 μm. Lower migrational velocities exist in this intermediate region and are a minimum for about 0.3-μm particles. This modified expression is

$$v_e = \left[(1 + 2 \times 10^{-5}/d)^2 + 2(1 + 2 \times 10^{-5}/d)^{-1} \right] \left(\frac{E_o E_p\, d^2 B}{4 \times 10^5} \right) \qquad (4.25b)$$

v_e is in cm/sec when d is in cm, E_o and E_p are in V/cm and B is in sec/g.

As an example, assume a high dielectric particle attains a saturation charge in an ESP operating with $E_o = E_p = 4000$ V/cm and $\epsilon_o = 8.85 \times 10^{-12}$ coulombs^2/(joule m) in air at SC, $\mu_g = 1.83 \times 10^{-4}$ g/(cm sec). Steady-state migration velocity for a 1.25 μm particle from Equation (4.24) is

$$v_e = \frac{(8.85 \times 10^{-12})(4000)^2(1.25 \times 10^{-4})}{(1.83 \times 10^{-4})} \times 10^5 = 9.7 \text{ cm/sec}$$

The factor 10^5 makes consistent units of these values.

For a 0.1 μm particle with a high relative dielectric constant and field strengths of 4000 V/cm, the migration velocity can be found using (4.25b) knowing from Table 2.1 that B is 1.66×10^8 sec/g:

$$v_e = \left[\left(1 + \frac{2 \times 10^{-5}}{1 \times 10^{-5}} \right)^2 + \left(\frac{2}{1 + \frac{2 \times 10^{-5}}{1 \times 10^{-5}}} \right) \right] \frac{(4000)^2(1 \times 10^{-5})^2(1.66 \times 10^8 \text{ sec/g})}{4 \times 10^5}$$

$$= 6.4 \text{ cm/sec}$$

A dust-free space would be created in a precipitator with laminar gas flow and no disturbances, as shown in Figure 4.6. This assumes all particles are small, the same size and of the same material. Under these conditions, all particles would be deposited within the distance

$$L = \frac{D\,\bar{U}}{2\,v_e} \qquad (4.26)$$

where D is the characteristic distance (e.g., diameter) of the system and $\bar{U}$ is gas bulk velocity.

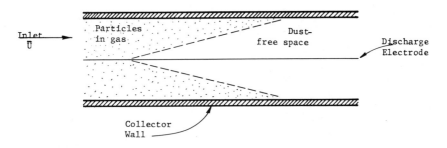

Figure 4.6   Schematic showing dust-free space with idealized laminar flow and no disturbances in an electrostatic precipitator.

Gas moves in turbulent flow in most precipitators with velocities of 0.9-1.5 m/sec. This is much greater than electrical velocities, which may be 0.15-0.30 m/sec. As a result of this continuous redistribution throughout the laminar free space, actual collection distance becomes much greater than L. As a result, collection efficiency in precipitators is often expressed using some form of the experimentally determined Anderson Equation as modified by Deutsch. The assumptions used are

1. Particles become fully charged as soon as they enter the system;
2. Particles are uniformly distributed through the cross section;
3. Particle migration velocity is not affected by gas velocity;
4. Particle motion is according to Stokes' Law;
5. Quasistationary electrical migration velocity;
6. No particle interaction;
7. No ion-neutral molecule collisions;
8. Particles remain on surface once they reach collector
9. No disturbances in gas flow.

The Deutsch modifications include a laminar dust layer, which exists near the collector surface. The Deutsch-Anderson equation for collection efficiency as given by White[15] for individual particle sizes is

$$\epsilon_i = 1 - \exp - \left(\frac{N_b}{\underline{N}} \frac{A_c\, v_e}{Q\, F} \right) \tag{4.27}$$

where N_b = particle concentration in laminar boundary region
$\underline{N}$ = average particle concentration
A_c = area of collector
Q = volumetric gas flow rate
F = a factor to account for unevenness of gas flow, equals ~ 1 for uniform flow and $\rightarrow 2$ for extreme cases.

The ratio $N_b/\overline{N}$ becomes more important as distance from inlet increases. The difficulty in this empirical equation is in obtaining values for each v_e and all the other terms in the exponential expression. The equation is useful for comparing similar or scale-up systems (see Section 3.6) knowing a measured overall efficiency for one unit.

An example of how Equation (4.27) can be used is given by expanding the calculations made with Equation (4.24) and (4.25b). These are given in Table 4.5. The sample particulate is fly ash from Figure 2.1 with the > 10 μm particles removed in a precleaner. The ESP charging and precipitating field strengths are 4000 V/cm, and long enough residence time is assumed for saturation charging of the larger particles. The fly ash is assumed to be in air at SC. The equations are not applicable for < 0.04 μm particles. Note that these migrational velocities are for the specific field strength noted. The minimal v_e and collection efficiency occurs for particles with diameters of about 0.3 μm. In the calculations given in Table 4.5, particle concentration differences are assumed negligible, and an even gas flow exists. Conversion of collector area to gas flow ratio to metric cgs units can be done by multiplying by 0.197 cm^2/(cm^3 sec) per 100 ft^2/1000 cfm. For example, for 1.25 μm particles fractional efficiency is

$$\epsilon_i = 1 - \exp\left[-\left(\frac{200\ \text{ft}^2}{1000\ \text{cfm}}\right)\frac{0.197\ \text{cm}^2/(\text{cm}^3\ \text{sec})}{100\ \text{ft}^2/1000\ \text{cfm}}\left(\frac{9.7\ \text{cm}}{\text{sec}}\right)\right]$$

$$= 0.961$$

A pseudo overall precipitation velocity to represent this sample under these conditions can be found by solving Equation (4.27) for v_e using overall efficiency. This is noted for the various systems in Table 4.5.

Other Surfaces

Collection of particles on other surfaces when one or the other or both are charged can be estimated by the potential and viscous flow theories. These, and others, provide approximate collection efficiency estimates for hybrid devices such as electrostatic augmentation of filters, scrubbers and precipitators. Various electrical forces present can be coulombic force between charged particle and charged collector; image force between charged particle and uncharged collector; image force between charged collector and uncharged particle; coulombic forces between charged particles of the same and opposite polarity; addition of electrical force field to any of the above; and magnetic (Lorentz) force field and charged particle.

The quasistationary migrational velocity (v_e), as shown by Equations (4.11) and (4.12) is equal to F_eB. Theoretical net attractive force for various

Table 4.5 Theoretical Fractional Collection Efficiency for Electrostatic Charging and Precipitation

Particle Diameter (μm)	Material Size Group (%)	Migrational Velocity (cm/sec)	Electrostatic Collection Efficiency Fraction for Collection Area to Gas Flow Ratio in ft^2/1000 cfm			
			200	300	400	500
<0.04	2		1.00	1.00	1.00	1.00
0.04	8	19.3	0.998	1.00	1.00	1.00
0.07	10	8.9	0.949	0.995	0.999	1.00
0.10	12	6.4	0.882	0.977	0.994	0.998
0.20	34	1.7	0.433	0.634	0.738	0.813
0.50	24	5.2	0.824	0.954	0.983	0.994
1.25	8.5	9.7	0.961	0.997	1.00	1.00
3.00	1.5	23.2	1.00	1.00	1.00	1.00
Weighted overall efficiency under these conditions			0.742	0.861	0.906	0.931
Pseudo overall precipitation velocity cm/sec			4.06	3.34	3.00	2.71

combinations of particle-surface types and charges can be estimated using the equations listed below. Subscript c represents collector and p represents particle.

Charged particle-uncharged collector—for conducting spherical collector[16]

$$F_e \cong \frac{4(n_p\ e)^2}{D_c^2} \tag{4.28}$$

and for conducting cylindrical collector[17]

$$F_e \cong \frac{(n_p\ e)^2}{D_c^2} \tag{4.29}$$

Multiply Equations (4.28) and (4.29) by $(\epsilon_c - 1)/(\epsilon_c + 1)$ for nonconducting collector surfaces.

Uncharged particle-charged collector—for conducting spherical or cylindrical collector[18]

$$F_e \cong \frac{\pi^2\ E^2\ d^3}{4\ D_c} \tag{4.30}$$

Table 4.6 is data calculated by Cooper and Rei.[18] It compares migration velocity of spherical particles and collecting surfaces for several of the conditions listed above. The notation $< v_a$ indicates less than sonic velocity of 3.31 x 10^4 cm/sec in air. For comparison, the same authors calculate migration velocity of conducting particles in an electrostatic field (as in an ESP) at 10,000 V/cm to be 45, 28.4, 58.8 and 155 cm/sec respectively for 0.1, 0.3, 1 and 3 μm spheres.

Table 4.6 Theoretical Terminal Velocities in cm/sec of Spherical Particles Approaching Collecting Surfaces—Maximum Field Strength is Assumed for Charging[18]

Conditions	Particle Diameter (μm)	Collector Diameter (μm)			
		0.1	1.0	10	100
Charged particle-	0.1	435	4.35	0.0435	4.35 x 10^{-4}
uncharged conduct-	0.3	855	8.55	0.0855	8.55 x 10^{-4}
ing sphere	1.0	1.83 x 10^4	183	1.83	0.183
collector	3.0	$< v_a$	4.2 x 10^3	42	0.420
Uncharged particle-	0.1		4.5	0.45	0.045
charged collector	0.3		22.1	2.21	0.221
	1.0		183	18.3	1.83
	3.0		1490	149	14.9

Values of single collector efficiencies can be related to migration velocities[19] by use of the approximation

$$\epsilon_i \cong \frac{v_e\, A_c}{\overline{U}\, A_x} \qquad (4.31)$$

where A_c is total collection area and A_x is collector cross section area perpendicular to gas flow which is moving at velocity $\overline{U}$. Collection efficiencies of charged particles on charged collectors of opposite charge[16] are for spheres

$$\epsilon_i = \frac{16\, q_p\, q_c\, B}{\overline{U}\, D_c{}^2} \qquad (4.32)$$

and for cylinders

$$\epsilon_i = \frac{4\pi\, q_p\, q_c\, B}{\overline{U}\, D_c} \qquad (4.33)$$

where q_p and q_c are charge on particle and collector, respectively.

It is obvious from inspection of migration velocities that electrical field augmentation can enhance collection if particle and surface charges are properly applied. When one member of the pair has no charge, the electrical force is one of attraction. When like charges are applied, a repulsive force is created that can hinder collection. Likewise, application of a force to move particles in opposition to the gas flow direction could retard collection while a force in the direction of motion could enhance it. Electrostatic collection is also enhanced by *increasing* the *charge* and by *reducing* the *collector* size.

Generally speaking, coulombic force of a charged particle in an electrical field is greater than image forces, which arise when either the particle or the collector are charged. Other phenomena occur as a result of particle charging, which can aid or hinder collection. Examples of this include magnetic influence on charged particles, agglomeration and mutual repulsion, scattering and deposition. Mutual repulsion of like-charged particles (or attraction if charges are opposite) produces a deposition force on a conducting sphere of

$$F_e \cong 2\, (n_p\, e)^2\, N_p\, (\pi/3)\, D_c \qquad (4.34)$$

where N_p is number concentration of particles. Other effects are discussed in appropriate sections.

4.3 MAGNETIC FORCE

There are two basic types of particles upon which the Lorentz or magnetic force can act when these particles are in a magnetic field. These are charged particles and magnetic particles. A charged particle with no intrinsic magnetic properties moving in a gas with a velocity U through a magnetic field will be acted on by a force (F_L) at right angles to both the direction of motion and direction of the field. The resultant Lorentzic force, expressed as a vector $\vec{F}_L$, is proportional to the magnetic field intensity (H), particle velocity in the moving gas stream and charge on the particle $q_p = n_p$ e. Expressed as the vector cross product, it is

$$\vec{F}_L = q_p \vec{U} \times \vec{H}/c \tag{4.35}$$

where H is the magnetic field intensity in oersteds and c is the speed of light (3×10^{10} cm/sec).

Magnetic field intensity of a solenoid can be estimated knowing

$$H = \frac{4\pi n I}{10} \tag{4.36}$$

where n is number of turns of conducting wire per cm and I is current in amperes.

Magnetic fields of up to about 6×10^3 oersted are possible. However, with optimum gas flow and magnetic field orientation, a comparison of the resultant force to maximum coulombic force shows that the coulombic force is about 10^4 times greater even when the particles (and gas) in the magnetic field are moving at the speed of sound. This makes magnetic force a poor particle-removal procedure.

Terminal magnetic drift velocity of a charged particle following Stokes' Law and Cunningham's correction with optimum flow and field orientation is

$$v_L \propto q_p \, U \, H \, B/C \tag{4.37}$$

Small magnetic particles in a magnetic field would tend to orient themselves in respect to their magnetic poles. Net magnetic attractive force could then be determined by algebraic resolution of the attraction and repulsion forces.

4.4 PHORETIC FORCES

The expression phoretic forces, as used here, designates forces that act on particles but are conveyed to the acting site by an indirect mechanism, as contrasted to the rather direct forces and fields discussed previously. These forces, in the form in which they act, are resolved in various levels of gas molecule energy in the molecules adjacent to the particle. The resultant force occurs when these molecules interact with the particles. Heat, light and molecule concentrations are the energy carriers producing thermophoresis, photophoresis, diffusiophoresis and Stephan flow. Phoretic forces are also known as radiometric forces.

Phoretic forces are very weak and become more significant as particle size decreases. In the past, these forces were often neglected because of this. The need to remove finer and finer particles makes it increasingly important to utilize these forces, to enhance and not hinder collection.

4.4.1 Diffusiophoresis and Stephan Flow

These forces are discussed first because they are perhaps the most important as far as practical particle collection mechanisms. Both forces are a type of diffusion force and often work together and can be considered as one. These forces are different from the particle thermal diffusion discussed in Section 2.4. Stephan flow is a result of the hydrodynamic flow of gas molecules normal to the surface of a volatile liquid necessary to maintain a uniform total pressure throughout the diffusing gases. Stephan flow causes a force to be exerted on the particles as a result of momentum exchange from gas molecules impacting on the particle as the molecules diffuse toward a condensing surface or away from an evaporating surface. It is an effect caused by the vapor molecules attempting to maintain a uniform distribution.

Diffusiophoresis is the net particle motion resulting from nonuniformities in the suspending gas composition. This includes Stephan flow as well as movement in the direction of the heavier or more concentrated gas molecule movement. It is the result of differences in molecular impacts on opposite sides of the particles.

Movement of fine particles in any wet scrubber can be influenced by diffusiophoresis. Stephan flow can cause more particles to be collected when the gas contains supersaturated vapor that can condense on the liquid droplets. Supersaturation of the vapor can result from cooling the gas below the dew point or by conditioning the gas by increasing the vapor concentration, as by spraying, for example. The latter procedure usually does both. The collecting droplet temperatures must be below the gas dew point also or the condensing vapor net flow will be away from them. Stephan flow can reduce the collection of particles by reversing the collection enhancement procedures, which occurs when the collecting droplets are heated and the gas is not saturated.

Diffusiophoresis theory is not significantly advanced to describe particle behavior entirely, but much work has been undertaken and some procedures have been developed. The velocity of the vapor, U_S, in Stephan flow from a binary gas-vapor system toward a carrier gas is

$$U_S = -\frac{D_{AB}}{p_g} \frac{dp}{dx} \tag{4.38}$$

where D_{AB} is diffusivity of the vapor in the gaseous medium and p_g is the partial pressure of the gas. Then

$$\frac{dp}{dx} = \frac{P^° - p_v}{\Delta x} \tag{4.39}$$

where $P^°$ is saturation vapor pressure of the droplet liquid at the temperature of the liquid, and Δx is distance from the droplet surface to the gas phase where vapor pressure equals p_v.

Diffusion force is the drag of the medium on the particle regardless of whether the motion is caused by diffusiophoretic actions or whether it is Brownian diffusion. This force is used to obtain values of diffusivity as given in Section 2.4. Diffusive force is discussed in Section 4.5 in an expanded manner. The sign on Equation (4.38) would be reversed for an evaporating liquid. Obviously several of these quantities cannot be obtained easily. In addition, this is the velocity of vapor molecules and not of the particle being moved by the Stephan flow.

Considering the combined effects of difference in diffusion molecular impacts and Stephan flow, Waldmann[20] developed an expression for the diffusiophoresis velocity, v_d, of very small particles in the Free Molecule Regime:

$$v_d = - \frac{P \sqrt{M_v} \, D_{AB} \, \nabla p_v}{p_g(p_v \sqrt{M_v} + p_g \sqrt{M_g})} \tag{4.40}$$

where P is total pressure and M_v and M_g are molecular weights of the vapor and gas. Grad p_v, ∇p_v, defines a vector field and physically represents the rate of change of p_v in all directions. Although this equation was developed for Free Molecule Regime particles, errors are $< 10\%$ for particles in the Transition Regime, which includes those up to about 0.4 μm in air.[21]

Diffusiophoretic deposition velocity of small particles (0.005-0.05 μm) from air in water vapor systems was studied by Goldsmith and May.[22] Using a nichrome aerosol at a number concentration of 3000-6000/ml, they reported deposition velocities equal to

$$v_d = - 1.9 \times 10^{-4} \frac{dp}{dx} \tag{4.41}$$

where v_d is in cm/sec and vapor pressure gradient is in mb/cm.

In the case where the particle itself is a condensing or evaporating droplet, the factors of mass and heat transfer complicate the particle motion behavior, which in turn is related to diffusiophoresis. These effects are most significant for Free Molecule Regime particles. Briefly, the mass of a droplet changes as condensation or evaporation occurs, and hence the momentum changes. Evaporation and condensation cause temperature changes as a result of both latent and sensible heat transfers. For example, a small evaporating droplet will cool, which in turn reduces the evaporative (and therefore the mass change) rate. Condensing molecules cause the reverse and, in addition, can exchange heat if they are at a different temperature. Rate of evaporation or condensation (ϕ) on a spherical droplet is related to the particle mass change:

$$\phi = 1/6 \; \pi \frac{d}{dt} (\rho_p \; d^3) \tag{4.42}$$

This in turn can be related to gas molecules using Avagadro's number (6.02×10^{23} molecules/g molecular weight). The number density, N_v, of evaporating molecules can be related to vapor pressure, p_v, using Boltzmann's ideal gas relation:

$$p_v = N_v \; k \; T \tag{4.43}$$

where T is the absolute temperature of the evaporating surface. Surface temperature of small evaporating droplets is reduced. Number density of particles leaving the surface is a function of surface temperature, composition and droplet size. An equilibrium net exchange rate would also depend on initial concentration of molecules present and the fate of these molecules.

4.4.2 Thermophoresis

A temperature gradient can result in motion of particles toward the colder region. Molecular mean thermal velocity can be determined using Equation (2.28), which shows $v_d = \sqrt{8kT/m}$ if molecule mass is m. This shows that in a uniform gaseous medium the molecules on the warmer side will move faster, striking the particles more often and also causing greater momentum exchange each time a particle collision occurs. The net force pushes the particle from the warmer area and can result in precipitation of the particle on a cold surface. This may cause the area close to a hot surface to be free of particles and, under certain conditions, appears as a dark space. This is called the black layer and increases in thickness as the temperature gradient increases.

For particles in the Free Molecule Regime, this net momentum exchange or thermophoretic force, F_T, is proportional to d^2. As estimated by Waldmann[20] it is

$$F_T \cong -\frac{P\lambda d^2}{T} \nabla T \tag{4.44}$$

where grad T (∇T) is the temperature gradient through which the force is observed. The gas molecule mean free path, λ, is at the temperature T of the particle. This force increases as gas molecule concentration goes up, as represented by pressure P.

Steady-state velocity of molecular regime spherical particles in a polyatomic gas under the influence of a thermophoretic force is given by Waldmann and Schmitt[23] as

$$v_T = \frac{6\,k\,\mu_g}{(8 + \pi\alpha)\,M_g\,P}\,\nabla T \tag{4.45}$$

which for ideal gases gives the velocity toward the colder region

$$v_T = -\frac{6\,\mu_g}{(8 + \pi\alpha)\,T\,\rho_g}\,\nabla T \tag{4.46}$$

where α is the coefficient of thermal reflection and represents the fraction of gas molecules reflected diffusely by the particle. Values of α are about 0.9 for liquids and particles with smooth surfaces, increasing to 1.0 for rough surfaces.

Large particles in the Continuum Regime are able to influence the gas molecules, and their behavior depends on heat conductivity of both the particle and the gas. A large particle acts as a barrier between the molecules colliding on the hot and cold sides. Epstein[24] predicted steady-state thermophoretic velocity of large particles in polyatomic gas using Maxwell's approximation when $k_p/k_g < 10$:

$$v_T = -\frac{3\,\mu_g\,k_g}{2\rho_g\,T\,(2k_g + k_p)}\,\nabla T \tag{4.47}$$

where k_g and k_p are thermal conductivities of gas and particle. When particle thermal conductivity is high, Brock's procedure[25] should be used:

$$v_T = -\frac{3c_T\,\lambda}{\rho_g\,d\,T}\,\nabla T \tag{4.48}$$

where c_T is a constant for the thermal transfer process defined by

$$c_T = \frac{15}{8}\,\frac{2 - \alpha_T}{\alpha_T} \tag{4.49}$$

The thermal accommodation coefficient, α_T, assumes accommodation of translational and internal energies are the same and depends on both the gaseous medium and the particles. These are best determined experimentally. Wachmann[26] gives values of α_T for various metals in air from 0.87-0.97, giving an average value of $c_T \cong 2.2$. Equation (4.47) is also suggested for use when particles are Slip Flow and Transition Regime size.

4.4.3 Photophoresis

Particles exposed to a source of light energy can be moved if the energy level is sufficiently high. The simplest explanation for movement is related to absorption of energy by the particles and subsequent heating of adjacent gas molecules. The effects then are similar to thermophoresis movement. However, there are factors that cause some particles to move toward the light source and others to move away. These include particle size, shape, transparency (transparent, translucent or opaque), color and refractive index. These factors plus the presence of other force fields, the intensity, color and type of light beam, and the gas pressure and composition all determine the particle motion pattern. Particle motion in a photophoretic field away from the light source (*i.e.*, in the direction of the light) is called positive, and motion toward the source is negative. This is consistent with thermophoresis and other phoretic forces.

Photophoretic movement may be completely irregular and have no directional tendency or there may be a net direction(s) influenced by external force(s). This leads to the general classification suggested by Preining:[27]

1) Irregular photophoresis—no marked direction of motion.
2) Simple photophoresis
   - light photophoresis (direction along light axis);
   - field photophoresis (gravito-, electro-, and magnetophotophoresis with motion in direction of respective field only).
3) Complex photophoresis—distinct motion in two or more directions.

Photophoretic motion is described as being like *enlarged* Brownian motion. Light striking a particle can be reflected and/or absorbed. The reflected light can be diffracted. The absorbed energy can increase the particle internal energy at the point of absorption and/or it can be refracted and transmitted, causing some local energy transfer to occur elsewhere in the particle. As a result, the particle is heated unevenly. This, plus irregularity of shape, results in uneven heating of local gas molecules and the ultimate random motion of the individual particles.

Estimation of photophoretic movement is extremely difficult because of the many complexities. Most phoretic studies are conducted in the presence of light energy of some wavelength for the convenience of being able to observe and measure particle behavior. Usually thermophoretic forces are not accounted for, which could explain some experimental variations. Atmospheric photophoresis could be an important factor in behavior of fine particles, especially at high elevations. Preining[28] gives a discussion on the mechanics of photophoretic motion.

4.5 DIFFUSION FORCE

Particle Brownian or thermal diffusivity is presented in Section 2.4 with the individual small particle mean diffusional velocity being given by Equation (2.33). A distinction is made here between particle thermal diffusion and diffusiophoresis and thermophoresis as defined in Sections 4.4.1 and 4.4.2. The latter exist because of a driving force gradient of molecule concentration or temperature. Yet the resulting force created by the drag of the particle moving through the gaseous medium is independent of whether the motion originates because of Brownian motion, diffusiophoresis or thermophoresis.

Diffusion force relationships vary with particle size regimes. Each particle in the Free Molecule Regime would create a drag force as a result of its random motion through the gaseous medium. The vector resolution of all these forces would be the diffusion force.

Motion of the larger particles in the Transition Regime are influenced by nearby gas-particle collisions, causing variations in the diffusion force. Studies have been carried out on particle motion in a binary gas for equimolar counter gaseous diffusion (ECD) and diffusion of one gas through a stagnant gas (SD). Thermal diffusion would more likely experience ECD conditions, and diffusiophoresis would more likely be SD. Brock[29] developed expressions for both ECD and SD diffusion forces. For ECD diffusion force of transition regime particles in binary gas A-B where A is the more concentrated,

$$F_d = \frac{1}{12} d^2 N_g (2\pi kT)^{1/2} (8 + \pi) D_{AB} \nabla x_A \left\{ (M_B^{1/2} - M_A^{1/2}) + \left[\frac{d_A M_A^{1/2}}{d_{AB}^{1/2}} - \left(\frac{2M_A M_B}{M_A + M_B} \right)^{1/2} \right] 0.156 \, d/\lambda' \right\} \qquad (4.50)$$

where N_g = gas molecule concentration
D_{AB} = molecular diffusivity of gas A in gas B
∇x_A = gradient of mole fraction of gas A
M = molecular weight of gas
d_A = diameter of gas A molecules
d_{AB} = ½($d_A + d_B$)

The mean free path of the binary gas mixture molecules is

$$\lambda' = 1/(2\pi N_g d_{AB}^2)^{1/2} \qquad (4.51)$$

For SD diffusion force of particles in a binary gas where A is the dilute gas,

$$F_d = \frac{1}{12}d^2 N_g (2\pi M_A kT)^{\frac{1}{2}}(8 + \pi)\frac{D_{AB}}{x_B}(\nabla x_A)\left\{1 - 0.0355[2M_A/(M_A + M_B)]^{\frac{1}{2}}d/\lambda'\right\} \tag{4.52}$$

where x_B is mole fraction of gas B.

An empirical expression for diffusion force of Slip Flow Regime particles[23] gives better estimates than the Continuum Regime Equation (2.25). For ECD in a binary gas A-B this is

$$F_d = 3\pi \mu_g d \ \sigma_{AB} D_{AB} \ \nabla x_A \tag{4.53}$$

and for SD it is

$$F_d = 3\pi \mu_g d (1 + \sigma_{AB} x_B) \frac{D_{AB}}{x_B} \nabla x_A \tag{4.54}$$

The term σ_{AB} is given by Equation (2.26).

Calvert *et al.*[30] developed an expression for the special case of particle collection by diffusion inside a bubble during the time the bubble rises through a liquid. This is

$$\epsilon_i = 1 - \exp - \left[\frac{12h}{\pi}\left(\frac{6 D_{PM}}{\pi D_b U_b}\right)^{\frac{1}{2}}\right] \tag{4.55}$$

where h is the height of the bubbling liquid, D_b is bubble diameter and U_b is bubble rise velocity.

4.6 ACOUSTIC FORCE

Sound energy is transmitted as longitudinal waves in contrast to the transverse electromagnetic light waves. As a wave, however, sound can be reflected, diffused and absorbed by particles. Sound is used in conjunction with other forces when collecting particles by causing them to coagulate or agglomerate. It is also used to enhance drop evaporation and condensation rates. Proper coupling of ultrasonic energy generators to collection devices is not easy and has led to numerous problems. Ultrasonics are mainly used in this work to eliminate noise pollution problems.

The effect of sonic energy on particles depends on whether the sound is confined as in a collection system so resonance and reflection can result in standing waves, or whether this sound is a free traveling wave as it can be in an unenclosed atmosphere. In a standing wave, nodes occur at a distance of one-half the sonic wavelength and antinodes are located midway between adjacent nodes. Stagnant gas molecules trapped in a standing wave will develop a circulatory pattern of motion from node to antinode. This does not occur in traveling waves. Particles suspended in the gas may move with the gas, drift or interact with each other, in addition to their basic thermal motion and motion caused by other forces. In this section low particle concentration is assumed, and interaction is neglected.

In addition to the above actions, particles in an acoustic field can, at various times, do some or all of the following:

1) vibrate to some degree with the gas.
2) drift in a translational direction from the longitudinal direction of gas movement,
3) circulate in the vibrating medium.

The motion of interest for particle removal is the drift of the particles. Particle drift velocities depend mainly on particle size and sound frequency and energy level. Drift velocities increase with frequency and are higher for larger particles. This drift can be explained by five mechanisms:

1) radiation pressure of the sound on the particles,
2) periodic change in viscosity of vibrating medium,
3) difference in phase of vibrating particles and the medium,
4) distortion of sound waves, and
5) asymmetry in a standing wave field.

These various forces and drift velocities are described by Mednikov[31] and are listed in Table 4.7. Symbols used in the table are:

ω = sound frequency

c_s = speed of sound in the gas

μ_f = flow around factor $\cong \omega\tau$ where $\omega\tau \ll 1$

$\overline{E}$ = absolute sound energy density (may be joules/c_s)

F_R = radiation pressure sonic force, etc.

v_R = radiation pressure particle drift velocity, etc.

x_1 = wavelength

x_2 = wavelength distance to nearest node

γ = ratio of specific heats of medium

b = $(2/d)(2 \mu_g/\rho_g \omega)$

ψ = phase shift angle of harmonic produced, equals $-\pi/2$ for second harmonic

h = relative amplitude of second harmonic produced

Note that in a standing wave the forces and particle drift velocities are zero at the nodes and antinodes, and are maximum half way between.

The predominant type of drift in undistorted traveling waves is due to difference in phase between particles and the medium. Acoustic radiation pressure drift velocities are negligible for undistorted traveling waves. The maximum velocity attainable by viscosity change plus phase difference is about 0.5 cm/sec. This is a function of particle size as μ_f is proportional to relaxation time.

Table 4.7 Forces and Drift Velocities of Suspended Particles in an Acoustic Field[31]

	In a Traveling Wave	In a Standing Wave
Radiation pressure	$F_R = 0.24 \left(\dfrac{\omega}{c_s}\right)^4 d^6 \mu_f^2 \bar{E}$ $v_R = \dfrac{0.046}{\mu_g}\left(\dfrac{\omega}{c_s}\right)^4 d^5 \mu_f^2 \bar{E}$	$F_R = 1.05 \left(\dfrac{\omega}{c_s}\right) d^3 \mu_f^2 \bar{E} \sin 2x_1 x_2$ $v_R = \dfrac{1}{9\mu_g}\left(\dfrac{\omega}{c_s}\right) d^2 \mu_f^2 \bar{E} \sin 2x_1 x_2$
Viscosity change	$F_\mu = \dfrac{3\pi}{2}\,\dfrac{(\gamma-1)\mu_g}{\rho_g c_s}\, d\, \mu_f^2 \bar{E}$ $v_\mu = \dfrac{(\gamma-1)}{2\,\rho_g c_s}\, \mu_f^2 \bar{E}$	
Vibration phase difference	$F_\phi = -3\pi\,\dfrac{\mu_g}{\rho_g c_s}\, d\, \mu_f^2 \bar{E}$ $v_\phi = -\dfrac{1}{\rho_g c_s}\, \mu_f^2 \bar{E}$	
Asymmetry		$F_a = \dfrac{\pi}{24}\left(\dfrac{\omega}{c_s}\right)\dfrac{d^3}{\mu_f}\left[\dfrac{9}{2}(b^2+b)\mu_f - (3+\dfrac{9}{2}b)\dfrac{1}{\mu_f}\right](\bar{E}\sin 2x_1 x_2)$ $v_a = \dfrac{1}{72\mu_g}\left(\dfrac{\omega}{c_s}\right)\dfrac{d^2}{\mu_f}\left[\dfrac{9}{2}(b^2+b)\mu_f - (3+\dfrac{9}{2}b)\dfrac{1}{\mu_f}\right](\bar{E}\sin 2x_1 x_2)$
Distortion	$F_h = -6\, h \sin\psi\, d^2\, \mu_f^2\, \bar{E}/4$ $v_h = -\dfrac{h\sin\psi}{2\pi\mu_g}\, d\, \mu_f^2\, \bar{E}$	

Distorted traveling wave drift velocities can reach velocities of 10 cm/sec and higher under proper conditions. Small particles in undistorted standing waves are affected mainly by asymmetry and attain velocities of $v_{a + \mu,max}$ $\cong$ 0.7 cm/sec. Large particles, however, are moved by acoustic radiation pressure at velocities of 40 cm/sec for 200 μm particles at 50,000 hertz/sec, for example.

In a distorted standing wave, small particles are moved mainly by asymmetry force, and large particles are influenced mainly by distortion force. For example, 10 μ particles can achieve a $v_{n, max} > 10$ cm/sec.

4.7 ADSORPTION FORCE

Diffusiophoretic forces and their effect on particle behavior have been discussed in Section 4.4.1. These forces exist when a gas molecule moves to or from a particle because of condensation or evaporation (Stephan flow), or because of net gas molecule momentum (diffusiophoresis). Adsorptive force should not be overlooked in this discussion because gas molecule movement in the vicinity of a particle surface may result from adsorption.

Adsorption is the taking up of molecules on the surface of a solid or liquid. The adsorbed molecules are held on the surface by electrochemical valence forces resulting from interaction with surface or near surface molecules. Adsorption is a common mechanism for removal of gas molecules and is well documented. The possible significance of gas molecule influence on suspended particle behavior should not be overlooked when the particle is highly carbonaceous, diatomaceous or a known good adsorbent. The resulting force, whether called an adsorption force or a diffusiophoretic force, can produce particle motion as discussed under diffusiophoresis. Adsorption of charged ions would obviously influence the electrostatic force effects.

Fine particles, with their large surface area-to-mass ratio, are good adsorbers. Volume of gas adsorbed per unit area is a function of partial pressure of gas on adsorbent, partial pressure of adsorbate, temperature, pressure and materials involved. Adsorption is also influenced by other conditions such as humidity and catalytic effects.

4.8 OTHER FORCES

It is obvious that all forces have not been discussed. In addition, many combinations of these forces can influence particle behavior. A significant one is centrifugal force, which has been used for a long time to remove particles from gas stream and is highly effective for large particles. It usually has only minor influence on particles less than about 3 μm in size. Centrifugal force effects are similar to gravitational effects and can be treated in the same manner by using the appropriate acceleration and by proper vector resolution of the particle motion. Some forces, such as adhesive force, result in particle-particle or particle-surface interaction (see Chapter 5).

Two interacting examples of particle behavior as a result of combinations of various forces are presented by Calvert *et al.*[30] The first is collection of particles inside rising bubbles by the *combination* of Brownian diffusion, diffusiophoresis and thermophoresis. This example assumes rigid, monodisperse particles < 0.3 μm in diameter inside a rising bubble of constant size, that the liquid is water and the gas is air and saturated with water vapor. Particle flux rate due to diffusion (N_{PD}) in units of mass per area time is estimated to be

$$N_{PD} = \frac{2N}{\pi} \left(\frac{6 D_{PM} v_b}{\pi} \frac{}{D_b} \right)^{\frac{1}{2}} \qquad (4.56)$$

where N = particle number concentration
v_b = bubble rise velocity
D_b = bubble diameter

Thermophoresis particle flux rate, N_{PT}, is obtained assuming the temperature gradient inside the bubble is linear in the radial direction and the aerosol temperature is the average of the gas temperature inside the bubble. T_i is temperature of inlet and T_e is temperature at exit when bubble leaves liquid. It is also assumed that circulation inside the bubble is such that no deposition occurs in the polar regions and occurs only $\pm$ 45° from the equator. From this

$$N_{PT} = \frac{\sqrt{2} C k_g \mu_g c_p v_b}{2(2k_g + k_p) h \, h'} \left(\frac{T_i - T_e}{T_i + T_e} \right) N \qquad (4.57)$$

where c_p = gas specific heat
h = bubble wall heat transfer coefficient
h' = height through which bubble rises

Particle diffusiophoresis flux rate, N_{PD}, inside the bubble is similarly shown to be

$$N_{PD} = \frac{\sqrt{2} M_v \, P \, N}{(p_i + p_e) \sqrt{M_v} + [2P - (p_i + p_e)] \sqrt{M_g}} \frac{D_b v_b}{6 \, h'} \frac{p_i - p_e}{[P - \frac{1}{2}(p_i + p_e)]} \qquad (4.58)$$

where P = total pressure
p_i = partial pressure of vapor inside bubble at inlet
p_e = partial pressure of vapor outside bubble at exit

From this, Calvert shows that fractional collection efficiency of a bubble on specific size particles inside the bubble by the combined forces of diffusion, thermophoresis and diffusiophoresis for temperature gradients $< 30°$ is

$$\epsilon_{D,T,D} = 1 - \exp \left[\left(\frac{12h'}{\pi D_b}\right)\left(\frac{6 D_{PM}}{\pi D_b v_b}\right)^{\frac{1}{2}} + \frac{3/2\sqrt{2}\, C\, k_g\, \mu_g\, c_p}{(2k_g + k_p)\, h\, D_b} \; \frac{T_i - T_e}{T_i + T_e} + \right.$$

$$\left. \frac{\sqrt{2}\, M_v\; P}{(p_i + p_e)\sqrt{M_v} + [2P - (p_i + p_e)]\sqrt{M_g}} \; \frac{p_i - p_e}{P - \frac{1}{2}(p_i + p_e)} \right] \quad (4.59)$$

The second example shows the effects of combined forces in the deposition of aerosol from a turbulent gas by inertia and Brownian diffusion. Note that additional combinations including gravitational and phoretic forces could be included if appropriate. In turbulent flow in a duct, the gas and suspended particles move in eddies, which result in a velocity toward the wall of the duct. This can affect particle removal by inertial impaction. Eddy diffusion decreases as the surface is approached because the drag of the dust surface reduces the gas flow rate near the walls. Eddy diffusion becomes negligible when a distance equal to approximately the particle stopping distance, x_s, is reached. Diffusion becomes significant in the quiet zone near the surface. Davies[32] shows this to be true when gas eddy diffusivity is $< 8 D_{PM}$.

Friction velocity, v_*, is defined in terms of the fluid pressure drop gradient down the pipe, $dP/d\ell$, as

$$v_* = \left[\frac{r_t}{2 \rho_g} \frac{dP}{d\ell}\right]^{\frac{1}{2}} \quad (4.60)$$

where r_t is radius of the tube. Pressure drop gradient can be estimated by

$$\frac{dP}{d\ell} = \frac{0.078 \, Re_f^{0.17} \, \rho_g \, \bar{U}^2}{r_t} \quad (4.61)$$

where $\bar{U}$ is bulk velocity of the gas. Dimensionless quantities, denoted by superscript $\sim$ are used here. Dimensionless velocity, $\tilde{v}$, is defined as

$$\tilde{v} = \frac{v}{v_*} \quad (4.62)$$

By a similar technique, a dimensionless length $\tilde{y}$ normal to the tube surface is

$$\tilde{y} = \frac{y \, v_* \, \rho_g}{\mu_g} \quad (4.63)$$

Analogous techniques provide dimensionless stopping distance, $\tilde{x}_s$, dimensionless particle radius, $\tilde{r}_p$, and dimensionless duct radius, $\tilde{r}_t$ (e.g., $\tilde{r}_t = r_t \, v_* \, \rho_g/\mu_g$).

Eddy diffusion of the turbulent gas produces an inertial deposition velocity (v_p) for particles < 0.3 μm. In terms of dimensionless velocity, $\tilde{v}_p$, this is

$$\tilde{v}_p = \frac{\tilde{v}}{I_{\tilde{y}}^{\tilde{r}_t}} \qquad (4.64)$$

where I is an integral representing one dimensional diffusion toward the surface from the center of the duct to a distance $\tilde{y} = \tilde{x}_s + \tilde{r}_p$. Some values for $I_{\tilde{y}}^{\tilde{r}_t}/1000$ are given in Table 4.8. The value of deposition velocity v_p

Table 4.8   Values of Integral $I_{\tilde{y}}^{\tilde{r}_t}/1000$ for Use in Equation (4.64); from Davies[32]

$\tilde{y}$	$\tilde{r}_t$			Ref
	100	1000	10,000	
0.04	547	547	547	10^4-10^6
0.06	211	211	211	10^4-10^6
0.08	108	108	108	10^4-10^6
0.1	65	65	65	10^4-10^6
0.2	14	14	14	10^4-10^6
0.5	2.10	2.13	2.54	10^4
0.5	2.08	2.09	2.12	10^5
0.5	2.07	2.07	2.08	10^6
1.0	0.550	0.582	0.991	10^4
1.0	0.542	0.551	0.602	10^5
1.0	0.535	0.538	0.544	10^6
2.0	0.162	0.194	0.603	10^4
2.0	0.157	0.166	0.217	10^5
2.0	0.152	0.155	0.162	10^6
5.0	0.0398	0.0716	0.481	10^4
5.0	0.0368	0.0460	0.0969	10^5
5.0	0.0341	0.0372	0.0436	10^6
10	0.0165	0.0483	0.458	10^4
10	0.0143	0.0236	0.0745	10^5
10	0.0125	0.0156	0.0220	10^6
50	0.00258	0.0343	0.443	10^4
50	0.00182	0.0110	0.0619	10^5
50	0.00129	0.00432	0.0108	10^6
100	–	0.0318	0.441	10^4
100	–	0.00923	0.0601	10^5
100	–	0.00303	0.00948	10^6

is equal to $\tilde{v}_p v_*$. The factor $\tilde{v}'$ is the dimensionless radial resolution of the root mean square turbulent velocity. It is dependent on particle size and is found from

$$\tilde{v}' = \frac{1}{2}\left\{(1 - \frac{\tilde{r}_p + 10}{\tilde{r}}) + \left[(1 - \frac{\tilde{r}_p + 10}{\tilde{r}})^2 + \frac{\tilde{r}_p}{\tilde{r}}\right]^{1/2}\right\} \qquad (4.65)$$

Near the surface of the tube, diffusion deposition occurs through the turbulent boundary layer, which has a dimensionless thickness no more than

$$\tilde{y}' \cong 10\left(\frac{8\,D_{PM}\,\rho_g}{\mu_g}\right)^{1/3} \qquad (4.66)$$

This is to say

$$\tilde{x}_s + \tilde{r}_p < 10\left(\frac{8\,D_{PM}\,\rho_g}{\mu_g}\right)^{1/3}$$

In this region, particle deposition rate due to diffusion only is

$$\tilde{v}_{pD} = \left[10\left(\frac{\mu_g}{D_{PM}\rho_g}\right)^{2/3}\frac{F + \mathbf{I}\,\tilde{r}_t}{10\left(\frac{8\,D_{PM}\rho_g}{\mu_g}\right)^{1/3}} + \frac{1}{\tilde{v}'}\right]^{-1} \qquad (4.67)$$

where

$$F = 0.7877 - 1/6\,\ell n\,\frac{(1 + \phi)^2}{1 - \phi + \phi^2} - \frac{1}{\sqrt{3}}\tan^{-1}\frac{2\phi - 1}{\sqrt{3}} \qquad (4.68)$$

The value of ϕ is

$$\phi = \frac{\tilde{x}_s + \tilde{r}_p}{10\left(\frac{D_{PM}\,\rho_g}{\mu_g}\right)^{1/3}} \qquad (4.69)$$

Depending upon where the particle is considered to originate, the appropriate equation must be used. Equation (4.67) would be used alone if the particle originates in the quiet zone near the wall. Equation (4.64) is used if the eddy diffusivity is such that diffusion can be neglected; otherwise use both. Davies[32] notes that diffusion through the turbulent boundary layer for particles in air near SC results in values for $\tilde{v}_p$ of about 1.9×10^{-2} for 0.0005 μm particles, 5.6×10^{-3} for 0.005 μm, 3.4×10^{-5} for 0.05 μm and 5.2×10^{-6} for 0.5 μm.

REFERENCES

1. Hidy, G. M. and J. R. Brock. *The Dynamics of Aerocolloidal Systems* (London: Pergamon Press, 1970).
2. Ranz, W. E. and J. B. Wong. "Impaction of Dust and Smoke Particles," *Ind. Eng. Chem.* 44:1371 (1952).
3. Lundgren, Dale A. "A Sampling Instrument for Determination of Particulate Composition, Concentration and Size Distribution Changes with Time," NAPCA Symposium on Advances in Instrumentation, Cincinnati (May 1969).
4. Goldshmid, Y. and S. Calvert. *AIChE J.* 9:352 (1963).
5. Taheri, M. and S. Calvert. *J. Air Poll. Control Assoc.* 3:129 (1960).
6. Kerker, M. *Aerosols and Atmospheric Chemistry,* G. M. Hidy, Ed. (New York: Academic Press, 1972), p. 3.
7. Adams, J. R. and R. G. Semonin. *Proceedings of the Precipitation Scavenging Meetings,* W. Slinn, Ed. (Washington, D.C.: U.S. Atomic Energy Commission, 1960).
8. Nichols, G. B. "The Electrostatic Precipitator Manual," The McIlvaine Co. (1976).
9. Billings, C. E. and J. Wilder. *Handbook of Fabric Filter Technology* NTIS, U.S. Dept. of Commerce, #P B 200 648 (1970).
10. Whitby, K. T. and B. Y. H. Liu. "The Electrical Behavior of Aerosols," In *Aerosol Science,* C. N. Davies, Ed. (New York: Academic Press, 1966).
11. Lowe, H. J. and D. H. Lucas. "The Physics of Electrostatic Precipitation," *Brit. J. Appl. Phys., London,* Suppl. No. 2 (1952).
12. Whitby, K. T., B. Y. H. Liu and C. M. Peterson. *J. Coll. Sci.* 20:585 (1965).
13. Lapple, C. E. "Electrostatic Phenomena with Particulates," In *Advances in Chemical Engineering,* T. B. Drew *et al.,* Eds., Vol. 8 (New York: Academic Press, 1970).
14. Cochet, R. "Charging Laws of Submicron Particles," *Colloq. Int. C N R S,* 102:331 (1960).
15. White, H. J. *Industrial Electrostatic Precipitation* (Reading, Mass.: Addison Wesley, 1963).
16. Kraemer, H. F. and H. F. Johnstone. "Collection of Aerosol Particles in Presence of Electrostatic Fields," *Ind. Eng. Chem.* 47:2426, as corrected in 48:812 (1956).
17. Lundgren, D. A. and K. T. Whitby. "Effect of Particle Electrostatic Charge on Filtration in Fibrous Filters," *Ind. Eng. Chem. Proc. Des. Dev.* 4:345 (1965).
18. Cooper, D. W. and M. T. Rei. "Evaluation of Electrostatic Augmentation for Fine Particle Control," U.S. EPA-600/2-76-055 (March 1976).
19. Zebel, G. "Deposition of Aerosol Flowing Past a Cylindrical Fibre in a Uniform Electric Field," *J. Coll. Sci.* 20:522 (1965).
20. Waldmann, L. *Naturforsch* 14A:589 (1959).
21. Waldmann, L. and K. H. Schmitt. *Naturforsch* 15A:843 (1960).
22. Goldsmith, P. and F. G. May. "Diffusio- and Thermophoresis in Water Vapour Systems," In *Aerosol Science,* C. N. Davies, Ed. (New York: Academic Press, 1966).

23. Waldmann, L. and K. H. Schmitt. "Thermophoresis and Diffusiophoresis of Aerosols," *Aerosol Science,* C. N. Davies, Ed. (New York: Academic Press, 1966).
24. Epstein, P. S. *Z. Phys.* 54:537 (1929).
25. Brock, J. R. *J. Coll. Sci.* 17:768 (1962).
26. Wachmann, H. *ARS Journal* 32:2 (1962)
27. Preining, O. *Staub* 39:45 (1955).
28. Preining, O. "Photophoresis," In *Aerosol Science,* C. N. Davies, Ed. (New York: Academic Press, 1966).
29. Brock, J. R. *J. Coll. Sci.* 17:768 (1962).
30. Calvert, S., J. Goldshmid, D. Leith and D. Mehta. *Scrubber Handbook* (San Diego, California: APT, Inc., 1972).
31. Mednikov, E. P. "Acoustic Coagulation and Precipitation of Aerosols," USSR Academy of Sciences Press in Moscow for the Institute of Combustible Minerals (1963).
32. Davies, C. N. "Deposition from Moving Aerosols," In *Aerosol Science* (New York: Academic Press, 1966).

CHAPTER 5

PARTICLE COLLECTION

5.1 IMPACTION AND INTERCEPTION

Collection of particles on surfaces by impaction and interception is introduced in Sections 3.5 and 4.1. The impaction mechanism has been noted as being the most common procedure for particle collection. Efficiency of impaction collectors is usually plotted as a function of "impaction parameter," which is a ratio of drag to viscous forces. A common impaction parameter, K_I, is the Stokes Number, which has been given as Equation (4.1):

$$K_I = \frac{d^2\, \rho_p\, v\, C}{9\, \mu_g\, D_c}$$

for Cunningham particles, where v is the relative velocity difference between particle and collector. Remember the caution urged in Section 4.1.1 when considering impaction parameters because of the number of common parameters in use. Note how each particular author defines the parameter. Collection efficiencies by impaction and interception mechanisms can be estimated by the procedures given in Sections 3.5 and 4.1, but *actual* collection also depends on whether the particles *remain* attached to the collecting surfaces. Reentrainment plus complications due to varying collector size and geometry sometimes make it necessary to use experimentally derived efficiency data rather than theoretical data.

5.2 FILTRATION

Particles can be removed from a gas stream by capture on or in a filter media. This media may consist of fibers woven into a fabric, a mat of fibers or layers of granular solids. Particles can be collected by any or all of the mechanisms of impaction, interception, Brownian or eddy diffusion, gravitational sedimentation and electrical attraction. Theoretical estimation of collection efficiency would require incorporation of all of these factors plus an assumption to account for reentrainment.

The major collection occurs when collected particles build up on the filter surface and bridge the pore openings, which are often larger than the particles themselves. The filter cake now acts as an efficient filtering medium. This process, in which the cake does the filtering, is called sieving. Billings[1] shows that typical filtration collection efficiency of a 0.3 μm particle varies for cloth filter type and condition. For example:

1) Light synthetic cloth equals 2% new, 13% cleaned, 65% with cake;
2) Heavy synthetic cloth equals 24% new, 66% cleaned, 75% with cake; and
3) Heavy natural cloth equals 39% new, 69% cleaned, 82% with cake.

Fabric filters may consist of yarns made from stranded fibers of a natural spun staple or a synthetic continuous monofilament. Figure 5.1 shows the cross-sectional appearance of several filter fibers. Not shown in this cross-sectional view are the hair-like appendages that project from woven and "dirty" fibers. Typical fiberglass and

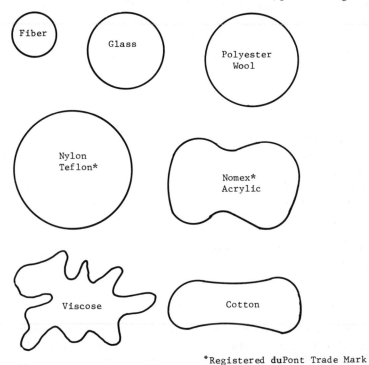

*Registered duPont Trade Mark

Figure 5.1

Figure 5.1 Cross-sectional appearance of filter fibers. (Courtesy the McIlvaine Co.[2])

dacron filaments are from about 6.5 to 30 μm in diameter, respectively, and are spun to form yarns ranging from 100-500 μm in diameter. The yarns are woven into a fabric cloth of specific orientation, such as shown in Figure 5.2, or are woven and either mechanically abraded (napped fabric)

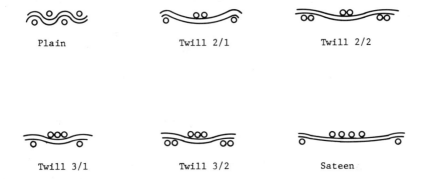

| Plain | Twill 2/1 | Twill 2/2 |

| Twill 3/1 | Twill 3/2 | Sateen |

Figure 5.2   Profile view of several types of woven filter cloth.

or needle-punched (felt). Hairs about the size of the filaments project from the fabric and are valuable for collecting particles because they effectively reduce the fabric pore size. Pore size is the distance of the opening between adjacent strands of yarn in the same layer of cloth and are from 50 to about 100 μm. Gas flow resistance or permeability varies with pore size, type of weave and treatment. The normal range for air is 10-150 cfm/ft^2 of fabric at a pressure differential of 0.5 inches of water.

Pressure drop across a filter depends on pressure drop of the porous medium plus pressure drop of the cake. The collection efficiencies noted above differ for a new and a cleaned (shaken, blown back) cloth because the latter retains some of the collected material in the pores. Pressure drop across a cleaned cloth increases for the first few cycles, then stabilizes. This stabilized pressure drop for a given cloth and type of particulate is the fabric medium pressure drop. Porous granular beds and other filtering media can act in a similar manner.

If the deposited particles build up so as to produce a uniform porous bed the pressure drop can be expressed by Darcy's Law:

$$\Delta P = -\frac{\mu_g v_g L}{K} \tag{5.1}$$

where L is the bed depth and K is permeability of the bed. Here, the velocity, v_g, is superficial velocity, which is the volumetric flow rate divided by the filter face area perpendicular to the gas flow. The value of K is best obtained empirically.

When the cake resistance does not differ significantly from the filtering medium resistance, it is not necessary to use two expressions (one for the cake and one for the filter). In the case of fabric filters, the cake resistance is typically of the same order of magnitude as the filter fabric residual resistance. The cake-fabric resistance due to deposition of particles over time interval, t, becomes

$$\Delta P_1 = (3.6 \times 10^3) K_1 v_g^2 C_0 t \qquad (5.2)$$

where v_g = gas velocity in cm/sec
$\quad\quad C_0$ = inlet particulate concentration, g/cm^3
$\quad\quad t$ = time, min
$\quad\quad K_1$ = cake-fabric resistance coefficient, cm $H_2O/[(g\ dust/cm^2)\ (cm/min)]$
$\quad\quad \Delta P_1$ = cake-fabric pressure drop, cm of H_2O

Williams[3] derived an expression for the cake-filter resistance coefficient using the Kozeny-Carman procedure for flow-through granular media at low-flow Reynolds numbers ($Re_f < 10$):

$$K_1 = (0.892 \times 10^{-2})\left(\frac{k}{g}\right)\left(\frac{\mu_g\ S^2}{\rho_p}\right)\left(\frac{1 - \epsilon}{\epsilon^3}\right) \qquad (5.3)$$

where   k = Kozeny-Carman coefficient, equals ~ 5 for a wide variety of fibrous and granular materials up to a porosity of ~ 0.8.
$\quad\quad \epsilon$ = porosity of void volume in cake layer, dimensionless fraction.
$\quad\quad S$ = ratio of particle surface area to particle volume in cake layer, cm^{-1} ($\cong 6/d$ for spheres barely touching).

Note that Equation (5.3) shows that as the particle size and therefore cake porosity decreases, K_1 increases, and because of this, pressure drop across the filter increases as particle size decreases.

Observed values of the cake-filter resistance coefficient do not always agree with those calculated by Equation (5.3). Data from Billings[1,4] are summarized in Table 5.1 for several particle diameters. The specific surface area is estimated assuming spherical particles at standard conditions.

Much of the variation between observed and calculated values of resistance coefficients results from the particle size and shape estimations used.

The terminal pressure drop ratio of "pressure drop in service prior to cleaning" to "cleaned filter pressure drop" is typically about 100:1. The terminal pressure drop ratio when new fabric pressure drop is used is about 200:1. Good fabrics can remain in service for about 10^4-10^7 cleaning cycles before they begin to deteriorate because of mechanical flexure and breakdown. When a break exceeds the capacity of the dust to bridge the gap the fabric should be replaced.

Table 5.1   Comparison of Calculated and Observed Dust-Fabric Filter Resistance
Coefficients for Industrial Cloth-Air Filters; from Billings[1,4]

Particle Size (μm)	S (cm^{-1})	Porosity ϵ	$\dfrac{1 - \epsilon}{\epsilon^3}$	Coefficient K_1 cm $H_2O/[(g\,dust/cm^2)(cm/min)]$	
				Calculated Using Equation (5.3)	From Observed Values and Equation (5.2)
0.1	6 x 10^5	0.25	48.0	7045	122
1	6 x 10^4	0.40	9.38	120	31
10	6 x 10^3	0.55	2.70	0.40	2.1
100	6 x 10^2	0.70	0.878	0.001	0.03

Physical and chemical resistances of some fibers are given in Table 5.2.[2] These characteristics are for comparative rating, and final selection should be based on actual operating test observations.

5.3 ELECTROSTATIC PRECIPITATION

Particles can be electrically charged and attracted to collector surfaces as discussed in Section 4.2.   Once on the surface, electrical resistivity of the collected particles becomes an important factor.   If this resistivity exceeds about 2×10^{10} ohm-cm, excessive sparking and reverse or back corona can occur, limiting the collection efficiency performance.   Particle electrical resistivity is inversely related to temperature and decreases in the presence of adsorbed liquid electrolytes.   Hall's curve as presented in the *Electrostatic Precipitator Manual*[5] is given as Figure 5.3.   This shows the effect of particle resistivity on current carrying capacity.   This curve was developed using dusts from actual operating plants and represents averages of numerous samples.

The electrostatic resistivity (ρ_R) of particles can be considered as the particle surface resistivity (ρ_s) and volume resistivity (ρ_v) acting in parallel. The electric field form of Ohm's Law gives

$$E = j\rho_R \tag{5.4}$$

where E is electrical field strength in volts per cm, j is amperes per cm^2 and ρ_R is in ohm cm.   Expressed as a sum of the two parallel resistances, an effective electrostatic resistivity value can be obtained:

$$\rho_R = \frac{\rho_s \rho_v}{\rho_s + \rho_v} \tag{5.5}$$

Table 5.2 Summary of Physical and Chemical Properties of Fibers (Courtesy The McIlvaine Co.[2])

Fabric	Approximate Year Introduced	Maximum Operating Temperature	Physical Resistance					Chemical Resistance				
			Dry Heat	Moist Heat	Abrasion	Shaking	Flexing	Mineral Acids	Organic Acids	Alkalies	Oxidizing Agents	Solvents
Cotton	–	180°F	G[a]	G	F[b]	G	G	P[c]	G	F	F	E[d]
Dacron®	1955	275°F	G	F	G	E	E	G	G	F	G	E
Orlon®	1953	275°F	G	G	G	G	E	G	G	F	G	E
Nylon	1946	225°F	G	G	E	E	E	P	F	G	F	E
Dynel®	1953	160°F	F	F	F	P-F	G	G	G	G	G	G
Polypropylene	1966	200°F	G	G	E	E	G	G	E	E	G	G
Creslan®	1953	275°F	G	G	G	G	E	E	E	E	G	E
Vycron®	1955	300°F	G	F	G	E	E	G	G	F	G	E
Nomex®	1964	400°F	E	E	E	E	E	P-F	E	G	G	E
Teflon®	1964	450°F	E	E	P-F	E	G	E	E	E	E	E
Wool	–	215°F	F	F	G	F	G	F	F	P	P	F
Glass	1939	550°F	E	E	P	P	F	E	E	G	E	E

aGood
bFair
cPoor
dExcellent

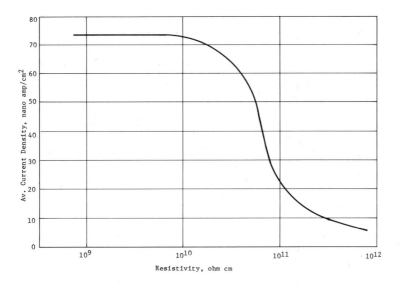

Figure 5.3   Hall's curve for full-scale ESPs of allowable current density versus particulate resistivity (Courtesy The McIlvaine Co.[5])

McLean[6] has developed theoretical expressions for particle resistivity including a compaction factor for both surface and volume compaction. The expression for surface resistivity is

$$\rho_s = A_s \exp(-p\, C_1\, e^{C_2/T}) \qquad (5.6)$$

and volume resistivity is

$$\rho_v = A_v \exp(E_a/kT) \qquad (5.7)$$

where p = vapor pressure, mm Hg
   C_1, C_2 = constants for the specific material
   A_s, A_v = constants dependent on compaction and nature of the particulates
   E_a = electron activation energy
   k = Boltzmann's constant
   T = absolute temperature

Values of the various constants can be determined knowing experimental resistivity values for a given dust. For example, surface resistivity dominates at higher temperatures and volume resistivity dominates at low temperatures, as shown in Figure 5.4 for a fly ash sample under two different water vapor pressures, p_1 and p_2. Obtain high temperature resistivity measurements for two different vapor pressures and solve Equation (5.6) to find values of

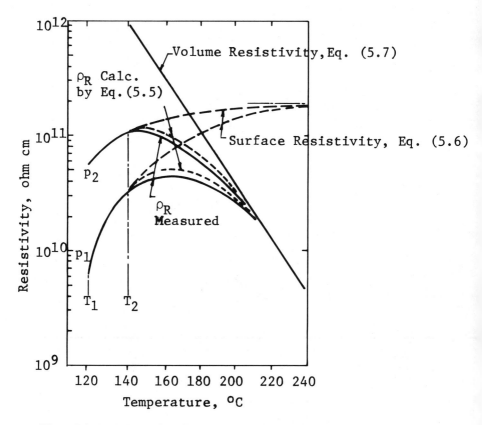

Figure 5.4   Resistivity curves for a typical fly ash sample at water vapor pressure, p_1 = 164 mm Hg and p_2 = 58 mm Hg.[6]

A_s and C_3, where C_3 replaces the expression $C_1 e^{C_2/T}$. Using experimental high temperature resistivity values at another temperature and some pressure, repeat the calculations to obtain C_1 and C_2. A similar procedure at low temperatures can be used to obtain values for the constants in Equation (5.7). Knowing these values, ρ_s and ρ_v can be determined giving ρ_R under any desired conditions. Typical values of these constants for fly ash are

$A_s \cong 2 \times 10^{11}$ ohm cm

$C_1 \cong 5 \times 10^{-8}$/mm Hg

$C_2 \cong 5080/^{\circ}$K

$A_v \cong 12$ ohm cm

$E_a \cong 0.9$ electron volts

Particles deposited electrostatically on a collector surface give off their electrons. As the dust layer builds up, the deposited particles must carry the charge from the newly deposited particles to the collector surface. Highly resistant dusts cause large electrical fields to build up. Localized field strengths build up and finally can exceed the electrical breakdown strength of the gas in the dust layer, which is about 10^4 volts/cm. The corona formed in this manner is called a back corona and the ions formed can recharge or discharge charged particles, interfering with their electrical drift toward the collection surface. A second detrimental effect is the effective reduction of the collecting field strength.

Collected particles must possess an adhesive force capable of holding them on the collected dust or they will be reentrained. Particles can be reentrained as a result of high gas velocity, by being dislodged by an entering particle or by electrostatic repulsion due to back corona action. Most particles are capable of withstanding gas velocities up to about 300 cm/sec before being swept away but newer, high-efficiency systems operate in the 60-120 cm/sec range. Fast-moving particles arriving at the collection surface can strike and dislodge collected particles. These ejected particles can, in turn, bounce on the surface causing craters and dislodging more particles. This is called "saltation."

To remove the built-up deposits of particles, the collection surface can be rapped to knock the cake free. The acceleration of the rapping (a) must exceed

$$a > \frac{T}{m/A} \tag{5.8}$$

where T is dust tensile strength, and m/A is mass per unit area. Sproull[7] found that it requires an acceleration of about 3×10^4 cm/sec^2 to remove 90% of fly ash with a tensile strength of about 10^3 dynes/cm^2.

5.4 COLLECTION DEVICES

This section covers briefly several general particle collection systems as they are related to fine particle behavior. It is not intended to discuss design, operation and maintenance of devices. If this is required, the reader is referred to other publications.[1,2,5,8-10]

Particles are contained within any collection device for some finite period of time as the gases and particles move through. This time, called residence time, τ_{res}, is established by the gas volumetric flow rate and equipment physical geometrical size and arrangement. Residence time is an *average* characteristic time unless true plug flow conditions exist with no backmixing. In general, collection time must be $< \tau_{res}$ for the particle to be removed, so this becomes a critical or limiting time. Average residence time is

$$\tau_{res} = \frac{V}{Q} \qquad (5.9)$$

where V is the active volume of the collection device and Q is volumetric flow rate.

5.4.1 Impactors

Collection by impaction depends on how the collection device is constructed and operated and the drag-to-viscous force ratio known as the impaction parameter, $K_I = (d^2 \rho_p v C)/(9 \mu_g D_c)$. This is summarized by the data calculated by Golovin and Putnam,[11,12] as presented in Figure 5.5. Pneumatic atomization of a liquid stream to produce droplets for collection of particles is a common technique. Venturi collectors are one type of device that makes use of this. Collection in a venturi could take place according to Figure 5.5 and the drop shape assumed; however, the wettability of the particle can cause a deviation from the appropriate curve.

Inertial collection of individual particles in a device such as a venturi scrubber can be controlled by the operating parameters included in K_I. If one assumes that the venturi scrubber has been designed properly, adequate liquid is available to completely cover the cross section of the device with collection droplets, and gas velocities are above the minimum critical velocity for atomization, pressure drop can then be related to particle overall collection efficiency.

Hesketh[13] gives a simplified equation for obtaining approximate venturi scrubber pressure drop:

$$\Delta P \cong \frac{v_t^2 \rho_g A^{0.133} L^{0.78}}{3870} \qquad (5.10)$$

where ΔP = pressure drop, cm H_2O
$\quad$ v_t = throat velocity of gas and particles, cm/sec
$\quad$ A = throat cross section area, cm^2
$\quad$ L = liquid-to-gas ratio, ℓ/m^3

Fractional collection efficiency for fine particles, $\epsilon_{<3}$, is

$$\epsilon_{<3} = (1 - C_f/C_o) \qquad (5.11)$$

where C_f and C_o are final and initial particle mass concentration respectively. Hesketh shows that in a venturi scrubber this ratio equals

$$C_f/C_o = 13.2 \, \Delta P^{-1.43} \qquad (5.12)$$

The ratio C_f/C_o is also called penetration, Pt. Venturis at reasonable pressure drop are essentially 100% efficient in the collection of larger particles so

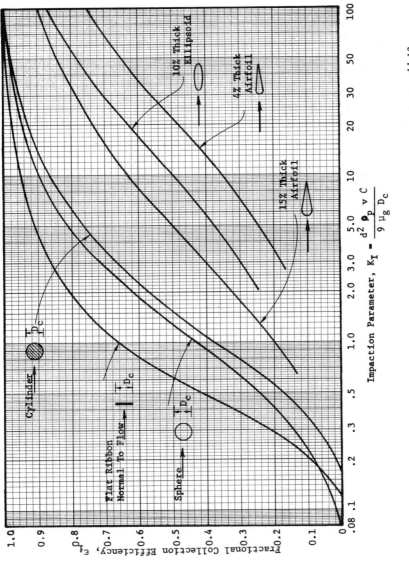

Figure 5.5 Estimated impaction efficiency as a function of inertial parameter for various targets.[11,12]

percent overall efficiency, $\overline{E}_o$, can be determined by

$$\overline{E}_o = \% \text{ large} + (\epsilon_{<3})\% \text{ fine} \tag{5.13}$$

where % large and % fine are the percent of large and percent of fine particles by mass distribution as determined for example from a log-probability plot. Efficiency of orifice scrubbers can be determined using Equation (5.12) if ΔP is set equal to two times actual pressure drop of orifice scrubber.

Note that the above procedure is empirical and does not attempt to specify the collection droplet shape configuration. It also does not account for wettability variations between various particles. Calvert[8] accounts for wettability using a factor f, which usually varies from 0.25-0.5. Cooper et al.[14] provide a corrected version of this technique, which is shown as Figure 5.6. This plots calculated penetration against venturi scrubber pressure drop for various values of f and various particle aerodynamic diameters.

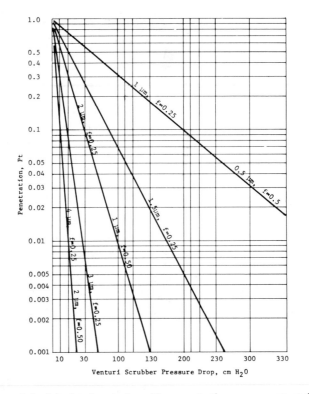

Figure 5.6 Calculated venturi scrubber penetration versus pressure drop as a function of particle aerodynamic diameter and f factor.

Collection by impaction using jets and orifices and multiple stages is covered in Chapter 6.

It should be apparent by now from the implications made above that particles can be removed by impaction only if they are within a specific interception area in regard to the collecting surface and have an appropriate drag-to-viscous force ratio. This results in a characteristic particle collection time called scrubber collection time, τ_{SC}. It is desirable that this time be as short as possible. A wet scrubber cannot be effective unless $\tau_{SC} < \tau_{res}$.

Particles within the area πy^2, where y is defined by Figure 5.7, could be expected to be collected by impaction on spherical droplets with a projected area of $\pi d^2/4$. Walton and Woodcock[15] suggest the following approximation can be used when $K_I > 0.7$:

$$y \cong \frac{D_c}{2(1 + \frac{0.7}{K_I})} \qquad (5.14)$$

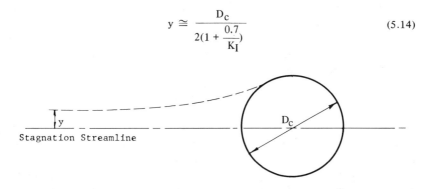

Figure 5.7 Particles approaching a spherical collection surface may be captured by impaction only if they are within the cross section πy^2.

A single droplet collector cleans a gas volume of $\pi v y^2/sec$ where v is the relative velocity difference between the particle and the droplet. With a particle number concentration of N, one droplet removes $N\pi v y^2$ particles per second and with a collecting droplet concentration per unit volume of N_D, the number of particles removed per unit time is $N_D N\pi v y^2$ or

$$\frac{dN}{dt} = N_D N \pi v y^2 = \frac{N}{\frac{1}{N_D \pi v y^2}} \qquad (5.15)$$

Equation (5.15) shows that the characteristic scrubber cleaning time, τ_{sc}, is the denominator of the right side, which can be written using Equation (5.14) as

$$\tau_{sc} = \frac{4(1 + \frac{0.7}{K_I})^2}{\pi v N_D D_c^2} \qquad (5.16)$$

Melcher and Sachar[16] show that for fine spherical particles in air near normal conditions, Equation (5.16) can be approximated using the Cunningham correction:

$$\tau_{SC} \cong \frac{48\,\mu_g^2}{v^3\,N_D d^4\,\rho_p^2\,C^2} \tag{5.17}$$

A 1 μm particle would have a τ_{SC} of about 0.13 sec when $v = 1000$ cm/sec, $N_D = 10^3$ drops/cm^3 and $\rho_p = 1$ in air at SC. For a 0.3 μm particle $\tau_{SC} \cong$ 0.82 seconds under the same conditions.

Another characteristic time in wet impaction systems is the scrubbing life time, τ_{SL}, of the droplet collectors. In a spray scrubber, droplets can decelerate to the gas velocity and in a venturi scrubber they accelerate to the gas velocity. Relative velocity difference is a very important parameter to the collection as evidenced by the presence of v^3 in Equation (5.17).

The tendency of a droplet to approach the gas velocity can be represented by equating droplet inertial force (mass x acceleration) to Stokes viscous drag force:

$$(1/6)\,\pi\,\rho_c\,D_c^3\,\frac{dv}{dt} \cong 3\,\pi\,D_c\,\mu_g\,v \tag{5.18}$$

where D_c is diameter of the droplet and ρ_c is density of the droplet. In impaction collectors, droplets are often 50 μm and larger and v is 10 m/sec giving $Re_p \sim 10\text{-}40$ so this is only approximate. Droplet interaction is assumed negligible, and droplet concentration is considered small enough so that momentum transfer between gas and droplets can be neglected. Rearranging this equation, the characteristic time is

$$\tau_{SL} \cong \frac{\rho_c\,D_c^2}{18\,\mu_g} \tag{5.19}$$

Droplet scrubbing life time is often a particle collection limiting factor. As droplet scrubbing life time falls below residence time, more droplets must be supplied or the collection efficiency becomes zero. This can be accomplished, for example, by injecting a series of sprays within a tower to subject the gas to numerous scrubbing stages. As τ_{SL} approaches the value of τ_{res}, the scrubber becomes a better gas absorber.

The relationship between effective scrubber collecting and droplet life time must be $\tau_{SC} \ll \tau_{SL}$. Using Equations (5.17) and (5.19) and solving for the lower limit of particle size that can be effectively cleaned by inertial scrubbing gives

$$d \gg 2\left[\frac{54\,\mu_g^3}{v^3\,N_D\,\rho_p^2\,\rho_c\,D_c^2\,C^2}\right]^{1/4} \tag{5.20}$$

where C is calculated using particle size.

Field test data reported by Abbott and Drehmel[17] on scrubber efficiencies show that the efficiencies of these devices decrease markedly as particle size decreases below about 1 μm (Table 5.3). No mention is made whether positive or negative phoretic forces are present.

Table 5.3 Field Tests on Scrubbers[17]

Particulate Source	Scrubber Type	ΔP (cm H$_2$O)	T ($^\circ$C)	Percent Efficiency for Particle Diameter (μm)			
				0.3	0.5	1	2
Coal-fired boiler	Chemico Venturi (580,000 actual m^3/hr)	25.5	165	7	47	90	99+
Coal-fired boiler	UOP TCA (640,000 actual m^3/hr)	30.5	145	10	70	90	98
Iron-melting cupola	Nat. Dust Collector Venturi-Rod (72,000 actual m^3/hr)	272	90	–	93-97	98.6-99.6	99.9+

5.4.2 Filters

In contrast to wet scrubbers, fabric filters must normally operate dry, and in most cases must be operated above the dew point of the vapor to reduce build-up due to caking of hygroscopic dusts. Some filters, such as the gravel bed filters, operate wet with the liquid used to flush off the deposited cake. Fabric filters are cleaned by shaking, blow back of gases or combinations of these. This means that the actual filtering operation is semicontinuous because of the cleaning interruptions.

Filter collection mechanisms include inertia, interception, sedimentation, diffusion and electrostatics. Collection efficiency is influenced by properties of the particulates, filter and cake. Collection efficiencies of filters can be very high but particles 0.1-1 μm in size are most difficult to remove because they are not effectively collected by the dominant mechanisms of impaction and diffusion. The deposition of larger particles is due mainly to inertia. As the cake builds up in a multi-layer filter the fabric openings close, and the collection mechanism changes more to interception, which in this case becomes a sieving action. Shallow fabric surfaces form a sealant dust cake quicker than napped surfaces and high permeability fabrics are less efficient when clean.

Microphotographs of particle capture on fabric filters show how deposition occurs. Particles collect mainly on previously deposited particles, and higher filtering velocities tend to produce more compact aggregates, which are closer to the fiber. Collection occurs on the front, back and sides of the fiber. The projecting aggregates bend in the gas flow and at higher velocities become detached when aggregates exceed about 10 particles. High-inlet dust concentrations aid in forming a cake quickly. Gas velocities increased above an optimum for each fabric-particle system lower filter efficiency. Normally, low velocities are < 3 cm/sec and high velocities are > 5 cm/sec.

Three distinct filtering regimes usually exist, each following $d\Delta P/dV = C_1(\Delta P)^{C_2}$. The constants are C_1 and C_2, and V is volume of gas filtered. Classical filtration theory applies when $C_2 = 0$ and therefore $(\Delta P)^{C_2} = 1$. Values of C_2 equal about 1.5 during the intermediate filtration and 2.0 when the cloth is clean and particles are only blocked or strained by the fabric.

A schematic representation of filter operation is given in Figure 5.8 where a constant gas flow rate and inlet dust concentration are assumed. Dust permeability is the slope of the curve or $\Delta W/\Delta S$ where W is mass deposited and S is drag. Permeability is constant during the interval B in Figure 5.8. The filter is cleaned at the end of B and operation is resumed at the start of A. Permeability increases with time in the early part of the cycle as the discontinuities on the filter surface are being repaired. Filter efficiencies are lowest during this period. This curve shape will vary

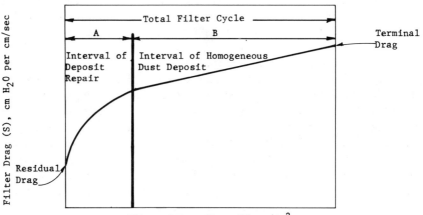

Figure 5.8 Schematic of fabric filter performance.

depending on the properties of particles, filter and cake, and operating conditions. In a pulse jet filter, Darcy's Law does not give the proper pressure drop-velocity relationship unless redeposition of the dust on adjacent bags is considered.[18]

In light of what has been presented, it should be apparent that overall filter collection efficiency is not a simple calculation. Efficiency depends on the particle, fabric and cake properties, gas velocity, length of filter cycle, type of cleaning, condition of fabric and humidity as well as combinations of all of these factors.

Initial collection efficiency for a filter can be estimated using the following: Inertial impaction is found using calculated values of impaction parameter $K_I = (d^2 \, \rho_p \, v \, C)/(9 \, \mu_g \, D_c)$ and choosing appropriate efficiencies from Figure 5.5. Diffusion fractional collection efficiency is approximately equal to $Pe^{-2/3}$ where the Peclet Number, Pe, is given by Equation (1.25) as $[D_c \, (v_p - v_c)]/D_{PM}$. Diffusion effects are negligible when $Re_p > 10^{-1}$ and are small for $10^{-4} < Re_p < 10^{-1}$. Direct interception fractional collection efficiency is approximately equal to R^2, where R is the direct interception group and equals the ratio d/D_c. This efficiency is approximate for both R and $Re_p < 1$. When particle size is $>$ filter pore size, the cake acts to sieve out the particles, especially those $> 10 \, \mu m$.

Sedimentation collection of particles in a fabric filter is usually neglected. In general, sedimentation aids collection when flow is downward and hinders collection when upward. A proposed general gravitational fractional collection efficiency expression is the ratio terminal settling velocity to local gas velocity with an appropriate sign for positive or negative effect.

Examples of actual filtration efficiencies are given by Billings et al.[19] for various systems and operating procedures. Collection of copper sulfate at light loadings (about 2 g/1000 m^3) at 5 cm^3/sec per cm^2 air-to-cloth ratio were 81.1% efficient by weight for a glass filter fabric, 71.3% for heavy wool felt, 63.7% for cotton sateen, 45.6% for napped orlon and 41.1% for light wool felt fabric. In another example with different particulate matter, the heavy felt was best and the cotton sateen the poorest. Use of a filter aid increased efficiencies of all to about 99%.

Preliminary U.S. EPA data[17] show that about 99.8-99.9 overall mass collection efficiency can be obtained on flue gas particulates. The fabric systems studied include glass/Teflon* (reverse air), nomex (reverse air) and glass/graphite (repressure air plus shake). These data are summarized in Table 5.4.

---

*Registered trademark of E. I. duPont de Nemours & Company, Inc., Wilmington, Delaware.

Table 5.4   Preliminary Data on Fabric Filter Systems[17]

Boiler Type	Air-to-Cloth Ratio (cm/sec)	Percent Overall Mass Efficiency	Percent Efficiency for Particle Diameter (μm)					Filter Type and Cleaning Method
			<0.3	0.5	1	2	4	
Pulverized	1:1	99.8	94.5	99.8	99.7	99.7	–	Glass/Teflon-reverse air
Coker (high-outlet dust loading)	1.5:1	99.8	99.9	99.2	99.4	99.6	99.6	Nomex-reverse air
Coker (low-outlet dust loading)	2.1:1	99.9	98+	98.9	98.3	99.3	99.3	Glass/graphite-repressure air plus shake

Billings[19] shows that increasing the number of raps in a shaker filter decreases the filter pressure drop, but also decreases efficiency.   Even so, 200 raps are often recommended per cleaning cycle.   More shakes than this result in an excessive expenditure of power and excessive filter fabric wear rate.   The same holds true for improving the filter by increasing the amount of reverse flush air used to blow back through the cloth to clean it.   By contrast, all other factors constant, increasing inlet particle concentration increases collection efficiency.

Filters have a characteristic residence time, τ_{res}.   The collection time equivalent to τ_{SC} would be a variable and would have to be determined for each collection mechanism as it dominates during each portion of the filter cycle.   The collector lifetime in this system would essentially be infinite.

5.4.3 Precipitators

A cut-away view of a typical single-stage electrostatic precipitator is given in Figure 5.9.   Electrostatic precipitator operation consists of (1) charging the particles, (2) migration of the particles to cause them to precipitate onto the collecting electrode surfaces, (3) discharging the collected particles and (4) removing the particles from the surfaces.   Actual collection efficiency during operation includes not only the factors discussed previously such as charging efficiency, electrical migration velocity, particle resistivity and collected material erosion and saltation, but includes operating factors such as gas sneakage, nonuniform particle concentration and reentrainment.   Actual Cottrell or single-stage precipitators operate at a field strength of 4-5 KV/cm, resulting in a collecting force 3000 times that of gravity for a saturated charged 1 μm particle.

Gas sneakage is the term used to account for the gas and particulates that pass through the nonelectrified regions of a precipitator.   These exist at the top of the unit where the electrical distribution, plate support and rapper systems are located and at the bottom sections where the collection hoppers are attached (Figure 5.9).

Use of baffles can force some of the gas to mix back into the electrified regions and use of multiple stages help reduce sneakage losses.   The effects of 0.1%, 1% and 10% sneakage are plotted in Figure 5.10 as collection efficiency versus reentrainment for a typical four-section ESP.[5]

Particulate concentration nonuniformities are more detrimental in an operating ESP than in other collection devices and should be minimized.   One way to do this consists of installing diffuser elements at the inlet of the ESP to help distribute the gas.   A diffuser could be a grid plate with rectangular openings over 40-60% of the area, or a perforated plate (*e.g.*, with 2 in. holes as shown in Figure 5.9) with 25-50% open area.   In addition, flow

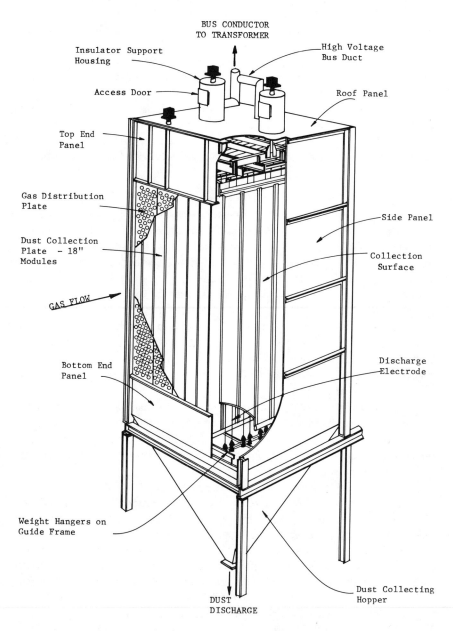

Figure 5.9 Typical precipitator cross section—single stage.

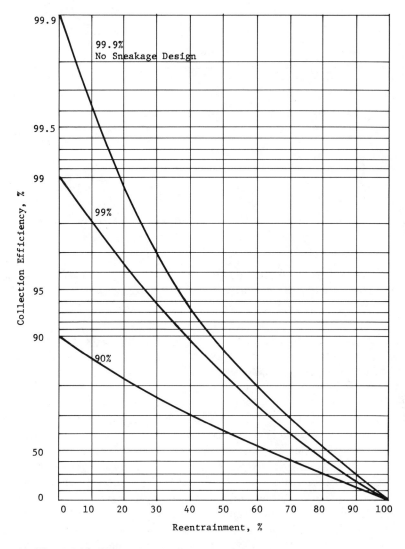

Figure 5.10 Effect of reentrainment and gas sneakage on a four-section electrostatic precipitator (Courtesy The McIlvaine Co.[5])

at turns and entrances should be straightened to maintain symmetry to reduce centrifugal stratification and/or eddy deposition of the particulates. Procedures for this are shown in Figure 5.11. In this sketch it is intended to imply that if needed, the two 30° angles would be in a plane perpendicular to the 90° angle.

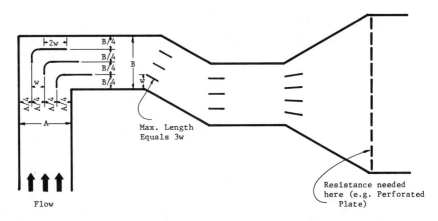

Figure 5.11   Techniques for reducing particulate concentration nonuniformities and reducing fallout accumulation.

Collected particles must be removed periodically. Rappers can be used to dislodge deposits of dry dusts, which then fall into the hoppers at the bottom. Sticky material can be removed by irrigating the collection surfaces with a liquid wash. Collected liquid particles would run off by gravity. The dry dusts cause the reentrainment problem. Nichols'[5] data show that more than 80% of this reentrainment emission occurs in the lower one-third of the precipitator. Rapping emissions summarized by Nichols indicate that these losses range from 6.5-53% of the total for the five installations reported (Table 5.5). Overall percent penetration is C_o/C_i times 100 or 100 minus percent overall efficiency. The pilot data indicate that if rapping is used, longer rap intervals seem to reduce reentrainment losses.

Nichols calculates the effects of rapping reentrainment for ESPs with one to four sections giving the estimated additional specific collection area needed to overcome these emission losses. The calculations as given in Table 5.6 are based on the model developed by Gooch[20] and are for coal-fired boiler fly ash with a precipitator operating at 20 nA/cm^2. This model shows that less additional collection area is required to achieve a specified overall efficiency as the number of ESP sections increase. Note that if reentrainment losses are less than about 2.3% per section no additional collection area is required in the four-section ESP while nearly 28% additional is required for the single-stage unit.

Field test data on ESPs[17] compares actual efficiencies of several operating systems. Increased efficiency is reported when dust resistivity is raised by either the presence of sulfur dioxide or by using elevated temperatures. These data are given in Table 5.7. Size and distribution of the particulates would vary from system to system, but the particulate sources were all coal-fired boilers.

Table 5.5 Percentage Contribution of Rapping Reentrainment to Total Emissions (Courtesy The McIlvaine Co.[5])

Installation No.	Comments	Type Rapper	Plate Acceleration (Gs) x,y,z axis	Rap Intervals (min)	Gas Velocity (m/sec)	Average Plate Current Densities[b] (mA/cm²)	Gas Temperature (°C)	SCA[a] $\left(\dfrac{m^2}{m^3/sec}\right)$	Total Penetration (%)	Penetration Due to Rapping Reentrainment (%)
1	SO$_3$ injection	Vibrator	3.7,1.6,2.9	4,6.5,6.5		7.6,10,4, 15.3,13.9	137		32.1	46
		Vibrator	3.5,1.8,2.6	8,13,13		11.2,11.2, 16.7,12.5	137		25.8	33
		Vibrator		4,6.5,6.5		11.2,11.2, 16.7,12.5	132		6.2	48
	SO$_3$ injection								17.4	
									3.2	
2	Pilot test	Drop hammer	11,16,15	12	0.87	23.3	122	33	11.4	53
	Pilot test	Drop hammer		32					7.6	32
	Pilot test	Drop hammer		52					6.1	18
	Pilot test	Drop hammer		150					6.9	25
	Pilot test	No rap							5.2	
3		Drop hammer		10,10,20,20	1.52	8.4,9.1,13.2	154	110	0.85	30
		Two plates		60,60		15.1,12.3				
4		Drop hammer		10,20,60	1.25	11.1,17.6,22.7	157	48	0.40	35
		Two plates				4.6,7.8,11.3				
5		Rotating drop hammer		6,6,12,12	1.83	14,24.6, 33.0,47.7	162	52	0.19	6.5

aSpecific collection area.
bFor each electrical field.

Table 5.6   Effect of Rapping Entrainment as a Function of Number of Sections and Constant Collection Efficiency Per Section (Courtesy The McIlvaine Co.[5])

Assumed Percent of Collected Material per Section Reentrained	Number of Sections	Penetration			Percent of Penetration due to Rapping Reentrainment (%)	Efficiency E_O (%)	Assumed Efficiency E_A Without Rapping Reentrainment (%)	Increase in E_O with No Rapping Reentrainment (%)	SCA Needed		Additional SCA [m²/(m³/sec)]	Additional SCA (%)
		Due to Rapping Reentrainment (%)	Without Rapping Reentrainment (%)	Total (%)					Without Rapping Reentrainment to Obtain E_O [m²/(m³/sec)]	With Rapping Reentrainment [m²/(m³/sec)]		
5.9	1	6.2	5.2	11.4	52	88.6	94.8	6.54	28.0	51.6	23.6	84.3
	2	1.02	0.27	1.29	79	98.7	99.73	1.03	62.1	105.4	43.3	69.7
	3	0.136	0.014	0.150	91	99.85	99.986	0.14	98.5	108.3	9.8	9.9
	4	0.0193	0.0007	0.020	97	99.98	99.9993	0.02				
2.3	1	2.23	5.20	7.43	30	92.57	94.80	2.35	40.4	51.6	11.2	27.7
	2	0.28	0.27	0.55	51	99.45	99.73	0.28	83.7	105.4	21.7	25.9
	3	0.027	0.014	0.041	66	99.959	99.986	0.03	108.5	108.3	0	0
	4	0.0023	0.0007	0.0030	77	99.997	99.9993	0.002				
0.97	1	0.90	5.20	6.1	15	93.9	94.8	0.95	46.3	51.6	5.3	11
	2	0.10	0.27	0.37	27	99.63	99.73	0.10	91.0	105.4	14.4	16
	3	0.009	0.014	0.023	39	99.97	99.986	0.016	108.3	108.3	0	0
	4	0.0007	0.0007	0.0014	50	99.9986	99.9993	0.001				

Table 5.7 Field Tests on Electrostatic Precipitators[17]

Type ESP	Specific Collection Area $(m^2/(m^3/sec))$	Tempera-ture (°C)	Percent Efficiency for Particle Diameter (μm)					Coal Sulfur Content
			0.1	0.5	1	2	Overall	
Cold side	56	150	98	95	97	98.6	99.6	Moderate
Cold side	47	155	99	99	99.6	99.9	99.8	High
Cold side	67	160	98	80	96	99	98.3	Low
Hot side	89	370	99.3	95	97	99.6	99+	No data

The preceding discussion shows how difficult it is to effectively apply theoretical collection efficiency expressions if actual operating loss mechanisms are not accounted for. In addition, the equations are developed assuming steady DC voltage. Actual operations usually use unfiltered rectified AC voltage. The resultant voltage pulsations affect the migration velocity, so an average velocity must be used. Keeping these points in mind, characteristic times can be defined to describe ESP operation. Use of Equation (4.12) and a distance s from the charging region and the collecting electrode gives a characteristic precipitator collection time, τ_{PC}, of

$$\tau_{PC} = \frac{s}{b\,E_W} \qquad (5.21)$$

where E_W is electrical field strength near the wall. According to Gauss' Law E_W is

$$E_W = \frac{N\,q\,s}{2\,\epsilon_0} \qquad (5.22)$$

where N is number concentration of particles, q is particle charge and ϵ_0 is dielectric constant of free space [8.85 x 10^{-14} coulomb cm/(cm^2 volt)]. ESP operation is not effective unless $\tau_{PC} \ll \tau_{res}$. A particle charging time could be expressed, but this is very small compared to τ_{PC}. The charge relaxation time required for collected particles to discharge after being collected is important and is related to the dust resistivity. However, this affects the value of E_W and should be included in the corrected values of E_W used.

5.4.4 Hybrid

Control devices that do not rely solely on mechanisms found in the conventional collectors are considered new concepts. There have been about 40 so-called "novel devices" proposed recently but most have been scrubbers with one or more unique features. The hybrid devices consisting of electrostatic augmentation have promise as submicron fine particle collectors. Both scrubbers and filters have been modified in this manner and are discussed in this section. Some are available commercially.

It should be mentioned that currently the "steam hydro scrubber" is an extremely effective novel device compared with those tested and does not use electrical augmentation. This collector uses waste heat to produce steam, which is then injected under high pressure into the throat of an aspirating-type venturi. Efficiencies on this unit by mass have been reported:[17]

Efficiency (%)	Particle Size (μm)
99.99	1
99.9	0.5
99.8	0.25
85	0.1
70	0.05
90	0.01

Mass overall efficiencies ranged from 99.84-99.9% on this device. Often high-efficiency devices reduce number emissions yet show little effect on overall mass efficiency, but this device is effective enough to show improved mass collection. Application of this equipment will probably be limited because of high energy consumption to processes where waste heat is available.

A packed bed filter is a relatively inefficient fine particle collector because of the loose packing and large pores. In conventional ESPs the interelectrode space is largely inactive for particle collection. This would be reduced in a hybrid charged filter bed where the entire surface of the medium could act as a collection surface. Particles would need to be charged prior to entering the bed. To date, only limited work has been done on this.

Electrostatically augmented mechanical collectors provide a substantial increase in collection efficiency. However, energy requirements are excessively high to make use of this technique at this time.

The hybrid of ESPs and wet scrubbers to produce charged drop scrubbers (CDS) has been studied most to date. There are numerous options available for CDS including charging the particles only, charging the collection droplets only, charging both particles and droplets the same or opposite polarity and charging both bipolar (a mixture of both charges to reduce self-precipitation). The most effective CDS technique has been to charge the particles one polarity and the drops the other. The space charge precipitator charges the particles only and relies on precipitation of particles and drops on the walls. A self-agglomerator device causes particle size to increase by agglomeration of small charged particles to enhance inertial collection. Other systems that use charged solids rather than liquid collectors include the electrofluidized and electropacked beds.

Schematically, a charged drop scrubbing system could be represented:

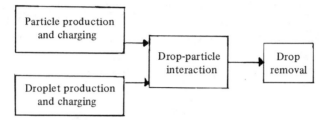

Physically, several of these boxes could be a single piece of equipment; conversely, a single box could represent several pieces of equipment. All except the droplet charging represent conventional facilities.

Droplet charging is accomplished by placing high-potential electrodes in the droplet formation area to induce charges or by use of a corona discharge to charge the drops by ion impaction. Induction charging occurs as the droplets are formed while ion impaction charges them after formation. Optimum droplet charge for best collection efficiency appears to be about

$$Q_{opt} \cong [12\,\pi\,\mu_g\,D_c\,\epsilon_0\,\frac{v}{\ell}\,N_{DO}]^{\frac{1}{2}} \qquad (5.23)$$

where ℓ is length of system, D_c is droplet diameter and N_{DO} is initial drop number density.

Three ubiquitous characteristic times are defined by Melcher and Sachar[16] to define the dynamics of systems of charged particles and droplets. Charged particles can interact with each other and either neutralize or self-precipitate (depending on charge polarities), resulting in a charged particle removal time, τ_p. Charged droplets can also interact with each other or otherwise lose their charge resulting in an effective charged drop life time, τ_D. Charged particles must travel to the droplets (or vice-versa) or walls. The time this takes depends on the space charge and is the collecting time, τ_C.

These characteristic times are defined for droplets and submicron particles:

$$\tau_p \equiv \frac{\epsilon_0}{b_p\,N} = \frac{\epsilon_0}{B_p\,q\,N} \qquad (5.24)$$

$$\tau_D \equiv \frac{\epsilon_0}{b_D\,N_D} = \frac{\epsilon_0}{B_D\,Q\,N_D} \qquad (5.25)$$

$$\tau_C \equiv \frac{\epsilon_o}{B_{Pq} \, Q \, N_D} = \frac{\epsilon_o}{b_P \, Q \, N_D} \tag{5.26}$$

where b is electric mobility, B is mechanical mobility, subscript P represents particles, subscript D represents droplets, q is particle charge, N is particle number concentration, Q is droplet charge and N_D is droplet number concentration.

The time for particle-droplet collision to occur is

$$\tau \equiv \frac{4 \, \pi \, \epsilon_o \, \ell_o{}^3}{3(q \, B_D + Q \, B_P)} \tag{5.27}$$

where ℓ_o is initial center-center spacing between particle and droplet which is $\gg (D_c + d)/2$. Three limiting cases exist: the first for bipolar charged particles only gives

$$\tau = \frac{\pi}{3} \, \tau_P \tag{5.28a}$$

The second occurs when particles are charged to one polarity and droplets to another:

$$\tau = \frac{\pi}{6} \, \tau_C \tag{5.28b}$$

Third, a system with bipolar charged droplets gives

$$\tau = \frac{\pi}{3} \, \tau_D \tag{5.28c}$$

A charged droplet collector using particles charged to one polarity and droplets charged to the other polarity (or charged bipolar) can improve the operation of a scrubber but would not compete favorably with a standard ESP because field strength limit determined by electrical breakdown places an upper limit on the drop charge, $N_D Q$. In comparison with a scrubber where it is assumed that τ_{SL} is adequately long and that both particles and droplets have saturation impaction charges (*i.e.*, $Q \cong 3 \, \pi \, \epsilon_o \, D_c{}^2 \, E_c$ and $q \cong 3 \, \pi \, \epsilon_o \, d^2 \, E_c$) the ratio of characteristic collecting times gives for submicron particles:

$$\frac{\tau_C}{\tau_{SC}} = \frac{v^3 \, \rho_p \, d^3 \, C}{452 \, \mu_g \, \epsilon_o \, D_c{}^2 \, E_c{}^2} \tag{5.29}$$

where E_c is charging field and v is relative velocity difference. For a CDS to be better than an inertial wet scrubber, this collecting ratio must be < 1.0. Equation (5.29) is significant as it shows that as the submicron particle size *decreases*, charged scrubbing effectiveness *increases* proportional to the cube of the particle size.

As an example, the collecting time ratio for 0.3 μm particles and 50 μm droplets charged to saturation can be determined. Assume v is 10 m/sec, both densities equal 1 g/cm^3, a linear charging field of 20 KV potential across a 2 cm gap and in air at SC. The resultant ratio is $\tau_C/\tau_{SC} = 0.23$. This is to say that under similar scrubbing conditions, the collecting time in a CDS may be, at best, about one-fourth that of a wet scrubber for 0.3 μm particles.

A comparison of CDS versus inertial scrubbing listing collection efficiency for 0.6 μm dioctyl phthalate particles captured by 50 μm water droplets gives inertial scrubbing as 25% efficient, CDS (particles charged, not droplets) 85-87% efficient, and CDS (particles and droplets charged) 92-95% efficient.

Self-precipitation of charged particles increases the effectiveness of CDS. As an example, Figure 5.12 shows collection of positively charged particles on negatively charged droplets. The dashed portion of the curve would be followed if no self-precipitation of particles occurred.

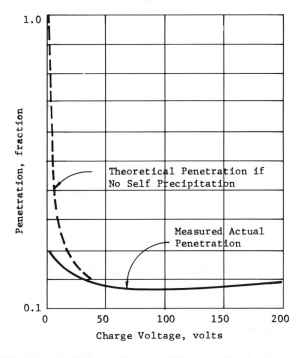

Figure 5.12 Typical efficiency of a charged drop scrubber showing the effect of self-precipitation of particles.[16]

5.5 MIST ELIMINATION

Suspended liquid particles in an exiting gas stream are considered to be mists. They can be a direct process release but they also usually accompany the discharge from a wet collection device. As a result, every wet collector must include a mist eliminator as an integral part of the system and often this limits the capacity of the system. In fact, mist eliminators (or mist separators) have been called the "Achilles heel" of scrubbing systems. If the mist droplets were pure water they would not be considered pollutants; but the mists can consist of suspended and dissolved solids and other corrosive chemicals.

Mists consisting of entrained liquid can cause other than air pollution problems. Loss of liquid from the system could be expensive. Scrubber entrainment can amount to 1% of the recycle liquid. Downstream system parts can be eroded or plugged by the substances present. This, in turn, can result in increased pressure drops, heat transfer resistance, system weight and vibrations.

Mist eliminators are usually separate devices, but could be simply large open spaces to permit coalescence and gravitational settling. Common entrainment eliminators include cyclonic separators, zigzag or louver baffles, knitted mesh, packed beds and other systems of vanes to provide a centrifugal separation force.

Capacity of separators is limited by reentrainment, which depends on gas velocity, entrained liquid concentration, liquid drainage rate and separator configuration and position. Gas velocity at which reentrainment occurs is called reentrainment velocity, and this is affected most by liquid drainage capacity.

Entrained droplets are usually relatively large so diffusional collection to remove these drops is not significant. This leaves the major separation mechanisms of: (a) inertial impaction, (b) sedimentation, (c) interception, (d) centrifugation, and (e) electrostatic precipitation. Electrostatic precipitation is being evaluated as a viable possibility and will probably be used more in the future on fine particle droplets. Centrifugation is a favorite procedure and can eliminate most of the large drops using high gas velocities.

The remaining mechanisms (a-c) are highly dependent on drop size, size distribution and inlet loading for separation efficiencies. Most separators operate at gas velocities of about 1-5 m/sec although some run as high as 12 m/sec. Sedimentation of the large drops entrained from a scrubber ranges from 0.1-2.0 m/sec so sedimentation is an important mechanism. The mechanism of interception is also significant for these size drops. Adequate gas flow is needed to achieve impaction, yet prevent excessive reentrainment. Reentrainment in straight ducts begins at velocities of about 18 m/sec, but because of design occurs at lower velocities in separators. Bell[21] reports that reentrainment in vertical louvers starts at about 5 m/sec; however, efficiency is low at this velocity. Increased velocity results in more reentrainment but efficiency increases to an optimum depending upon liquid loading and configuration of the specific collector.

Calvert[22] predicts values for maximum gas velocities to prevent excessive reentrainment for different separators. Reentrainment in zigzag (louver) baffles depends strongly on liquid drainage rate so orientation of the baffles is a variable. Best drainage is from a cross flow separator with verticle baffles and horizontal gas flow. In this arrangement, reentrainment becomes heavy when gas velocities exceed about 5 m/sec at liquid-to-gas ratios of 10^{-3} m^3/m^3. Gas velocities can be increased without excessive reentrainment at lower liquid-to-gas ratios. When orientated so the baffles are horizontal and gas flow is upwards, the reentrainment becomes excessive at gas velocities over 4 m/sec at 10^{-3} m^3/m^3. Maximum velocities for mesh are a function of liquid-to-gas ratio. At low liquid loadings (10^{-5} m^3/m^3) it is 5 m/sec and decreases with increased loading. For a cyclone, reentrainment was observed at velocities over about 40 m/sec.

For comparison, a form of the Souders-Brown Equation is suggested[23] to predict maximum allowable gas velocities in wire mesh separators to prevent reentrainment. In metric units of m/sec this velocity, v_g, is

$$v_g = 30.5 \, C_1 \sqrt{\frac{\rho_L - \rho_g}{\rho_g}} \tag{5.30}$$

where ρ_L and ρ_g are liquid and gas densities and C_1 is an empirical constant.

Pressure drops across separators do not seem to vary significantly with orientation. Amount of liquid entrainment has little effect on pressure drop except for knitted mesh. In the mesh separator, ΔP equals approximately

$$1.70^{\sqrt{L/A}} \, \Delta P_{dry}$$

where L is liquid rate in m^3/min, A is cross-sectional area in m^2 of mesh in liquid flow direction and ΔP_{dry} is pressure drop of dry mesh at that gas velocity. The most significant variable for a given system is gas velocity. In mesh units, for example, ΔP varies with gas velocity to the 1.65 power.

Overall efficiency of separators is considered to be $\geq 99.5\%$ when designed and operated properly. Depending on the size and size distribution of material to be collected, for example, a series of wire mesh pads can be installed to cause coalescing if needed as well as to do the collecting. Obviously this increases system pressure drop. The 99.5% minimum efficiency is a must for large commercial applications. Calvert's[22] wire mesh test units gave essentially 100% efficiency at gas velocities below about 5 m/sec at low entrainment loadings for typical scrubbing system size drops as noted. Efficiency dropped as liquid loading increased.

If the system is overloaded, as noted earlier, reentrainment occurs. Reentrainment mechanisms consist of (a) transition of liquid from separate liquid phase flow to separated-entrained flow, (b) rupture of bubbles, (c) creeping of liquid on the entrainment surface and (d) shattering of drops. The mass median drop diameter due to reentrainment as reported by Calvert[22] varies from 80-750 μm; however, studies in progress indicate the bulk of reentrained droplets may be < 10 μm.

The average drop diameter of separated-entrained flow droplets is reported as 250 μm, and size distribution is independent of duct dimensions. The amount of these droplets depends on gas velocity, liquid Reynolds number and liquid properties. Reentrainment in separators occurs at lower gas velocities than in straight tubes because gas jets are developed in separators and gas streams impinge onto the liquid at an angle. Therefore, reducing sharp angles reduces reentrainment and zigzag baffles inclined at 30° from horizontal gas flow direction should give less reentrainment than baffles inclined 45°.

Rupture of bubbles produces droplets < 40 μm and is reported to be insignificant in amount. This occurs mainly in plate towers, packed beds and mesh pads. Bubbles leaving the liquid phase rupture when the liquid film is about 0.1 μm thick. The bubble life is several 10^{-2} seconds, the burst takes several 10^{-6} seconds and the droplet formation requires several 10^{-3} seconds.

Creeping of liquid occurs when gas velocities are sufficient to drag the liquid along with the gas preventing the liquid from draining, e.g., by gravity. This can result in separated-entrained flow of droplets.

Shattering of drops, or pneumatic two-fluid atomization, results when gas velocities are sufficiently greater than those of the large liquid drops. In entrainment separators this produces relatively large droplets > 200 μm in diameter. Smaller drops are more likely to be carried away instead of being atomized to form droplets. This mechanism is common in zigzag baffles and also can occur where separated-entrained drops are present.

5.6 PRESSURE AND TEMPERATURE

Variations of pressure and temperature from standard conditions can influence the behavior of particles both directly and indirectly. Liquid particulate matter could be affected directly because of vapor pressure changes. Elevated temperatures can cause evaporation and reduced temperatures cause condensation. Section 4.4.1 notes how a direct effect can occur when Stephan flow causes particles to move because of the movement of gas molecules when vapor and liquid coexist in a gas stream. This movement is controlled by the temperature difference between the gas and liquid droplet phases. The inverse of these effects could occur when pressure is increased.

The indirect effect of pressure and temperature variations on particle behavior results because of changes in the physical properties of the gaseous medium. The properties of the gas affected most are viscosity and density, and at least one of these properties occurs in each particle behavior equation. Calculations for particles in the Cunningham Regime have to include an appropriately corrected slip factor that varies with temperature.

If pressure and temperature do not deviate greatly from SC, density can be estimated for most gases using the ideal gas law:

$$\rho_g = \frac{M\,P}{R\,T}$$

(5.31)

where   M = molecular weight of gas, g/g mole
        P = absolute pressure, atm
        T = absolute temperature, $^\circ$K
        R = universal gas constant, (82.05 atm cm^3)/(g mole $^\circ$K)

When temperature and pressure are significantly different from standard conditions so that ideal gas behavior cannot be assumed, then compressibility factors and/or departure tables or some other appropriate equation of state can be used.[24,25]

Viscosity of the gaseous medium must be carefully corrected for temperature variations. Remember, viscosity of gas increases with temperature (viscosity of liquid decreases with temperature). Tables of viscosity can be used if available. Values for air are given at 1 atmosphere in Figure 5.13.

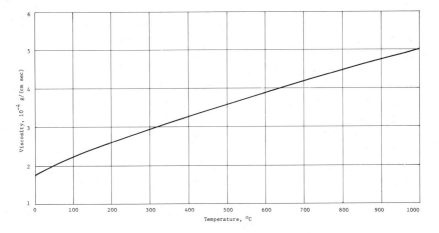

Figure 5.13 Viscosity of air at 1 atmosphere pressure.

Viscosities can also be calculated when necessary. The Chapman-Enskog Equation for low-density, nonpolar gases is

$$\mu_g = 2.6693 \times 10^{-5} \frac{\sqrt{MT}}{\sigma^2 \, \Omega_\mu}$$ (5.32)

where M is gas molecular weight, T is degrees Kelvin, Ω_μ is the viscosity function and σ is collision diameter (a Lennard Jones parameter). The value of Ω_μ is obtained knowing ϵ/K, another Lennard Jones parameter. These parameters are peculiar to each gas and must be estimated for mixtures. Tables for these parameters are available.[26]

Calvert[8] shows that increased temperature could result in increased thermal diffusion collection and decreased inertial, electrostatic and centrifugal collection; and increased pressure decreases collection by all mechanisms with a negligible effect on centrifugal collection. The effect on diffusiophoresis could be either depending on the magnitude of Stephan Flow.

5.7 INTERACTION OF PARTICLES

Particles in a gaseous medium are constantly in motion because of Brownian diffusion, system turbulence, eddy turbulence, evaporation, presence of gravitational, electrical and other force fields, gradients of concentration, pressure and temperature or other factors. As a result they continually contact each other and the adjacent system surfaces. The normal result is that fine particles stick to each other or the other surfaces, decreasing the number concentration of suspended particulate matter and changing the particle size distribution.

The hydrodynamic interaction effects of groups or clouds of particles and of adjacent surfaces on particles moving in a gas stream are discussed in Chapter 2. Hydrodynamic interaction occurs when the presence of an adjacent particle(s) causes the particle(s) to behave differently from single particle behavior. In contrast to the normal laws of physics, which state that two objects are attracted proportional to their mass and inversely as the square of the distance between them, fine particles may be either attracted or repelled by each other when no external forces are present. These forces vary with flow Reynolds number.

Fuchs[27] shows that the interacting force between two adjacent particles falling at low Reynolds number but $(Re_f > 0.1)$ is

$$F_s = \frac{3 \, \pi \, \mu_g \, d_1^3 \, d_2^3 \, v^2}{64 \, s^4} \left(\frac{3}{2} \cos 2\theta + \frac{1}{2}\right)$$ (5.33)

where s is center-to-center distance between the particles, θ is the angle between a line through the centers and the direction of motion, and subscripts 1 and 2 represent the two particles. When the particles fall parallel, side by side, there is a maximum attraction force. Falling one behind the other, the force is a maximum repulsive force. The force is of attraction when $\cos 2\theta < -1/3$ and repulsion when $> -1/3$.

Interacting particles falling vertically, one behind the other, have a faster terminal settling velocity, v_s', than the terminal velocity of an individual particle, v_s. The closer the particles become, the greater the interaction terminal velocity, v_s', becomes. For identical particles falling in this manner, the terminal settling velocity of the interacting particles is

$$v_s' = (1.6 \ d/s + 1) \ (v_s) \tag{5.34}$$

for values of d/s up to 1.0.

A Stokes' approximation can be used where the particles fall in line but have different diameters. This is

$$v_{s1}' = v_{s1} + \frac{3}{4} \frac{d_2}{s} \ v_{s2} \tag{5.35}$$

Equation (5.35) can be used for either particle by interchanging subscripts. Fuchs reports that this equation is good for $v\rho_g/\mu_g < 0.2$ but not accurate at small distances because observations show the first particle moves more slowly than predicted. This is because of the asymmetry setup as this particle moves through the gaseous medium.

5.7.1 Adhesion

Fine particles readily contact and adhere to solid or liquid surfaces. The resultant bonding or adhesive force can be very strong and, for individual particles, results from polar or other electrostatic forces. For example, the earth's gravitational force on a 50 μm quartz sphere is about 2×10^{-2} dynes. The gravitational attraction force between two of these spheres is only about 2×10^{-16} dynes, yet the separation force necessary to overcome the adhesion is about 0.5 dynes. This adhesive strength is similar in magnitude to the molecular strength of the quartz. This is why it is sometimes difficult to remove material from a collection surface if the initial layer adheres strongly to the surface.

At least six aspects of particle adhesion can be considered. These include adhesive force for individual particles, electrostatic charge, relative humidity, particle shape, surface roughness and group effects. Krupp[28] suggests that for individual spherical particles when no external forces are present that: particle and substrate contact at one point of atomic dimensions; attraction forces develop in several quickly formed contact areas; and equilibrium between attraction forces and deformation results in a final adhesive area of finite size. The *initial* force holding the particle could be a stopping kinetic energy loss, gravitational or other force.

Krupp further suggests that separation requires force to cause recovery of interface deformation (partial or complete). A nonsymmetrical separation force causes individual contacts to break separately. The particle is free only after the last contact is broken.

The attractive forces can be considered as long-range, short-range and interfacial reactions. The long-range attractive forces include van der Waals forces and electromagnetic and electrostatic forces. Short-range attractive forces are chemical bonds. Interfacial reactions come from diffusion and include dissolution or alloying.

Humidity and surface roughness can substantially influence adhesive force. Soluble substances could dissolve in high humidity conditions and increase either short-range or interfacial reaction forces. Presence of moisture due to condensation can introduce surface tension forces. Rough surfaces can result in small particles becoming embedded in fissures and larger particles having increased numbers of contact points.

Adhesive force of particles on a surface varies considerably for the same material. For example, consider 4 μm particles of iron on an iron substrate at room temperature and constant humidity. About half the 0.4 μm particles can be removed by a force of $< 4 \times 10^{-3}$ dynes while the final 2% require $> 30 \times 10^{-3}$ dynes. Adhesive force of iron particles on another surface could be more or less than this, but it would normally be greater for dissimilar substances.

Particle adhesive force varies with the factors mentioned including particle material and size, surface material and roughness, impact type, contact time and humidity. Even so, a first approximation of adhesive force, F_a, in dry air at standard conditions on a noncharged particle is

$$F_a \cong 10^{-3} d_p \qquad (5.36)$$

for removal of the first 50% of particles from a surface, and

$$F_a \cong 10^{-2} d_p \qquad (5.37)$$

for removing 98%. F_a is in dynes and d_p is particle diameter in μm.

Charged particles and surfaces would have a much higher adhesive force. An estimate of the magnitude of this force in air can be made using

$$F_a \cong \frac{n_p n_s}{4 \times 10^8 s^2} \qquad (5.38)$$

where n_p and n_s are number of electrons on particle and surface and s is separating distance in μm (usually particle radius).

The effect of relative humidity could be to produce a minute pool of liquid at the particle-surface interface and thereby increase adhesive force. This effect can be estimated by multiplying Equation (5.36) or (5.37) by the empirical relationship $[0.5 + (4.8 \times 10^{-3})\,(\%\ RH)]$ where RH is relative humidity. This applies for particles $> 20\ \mu m$ and for RH from 50-95%.

Surface shape can be accounted for by using the expression for 98% removal for two spheres:

$$F_a \cong 10^{-2}\,\frac{d_{p_1}\,d_{p_2}}{d_{p_1} + d_{p_2}} \tag{5.39}$$

and for a sphere-cylinder (e.g., as in fiber filter):

$$F_a \cong 10^{-2}\,d_p\,(1 + \frac{d_p}{D_c})^{-1} \tag{5.40}$$

Surface irregularities are called asperities and these are often $\sim 1\ \mu m$ or less in size. Adhesive force of a particle on a surface increases as asperity size decreases. An empirical approximation for asperity height, h, in the 0.01-10 μm range for 98% removal of 50-250 μm particles is

$$F_a \cong 10^{-2}\,d_p\,(1 + 10^2\,h/d)^{-4} \tag{5.41}$$

In the above equations, d, h and D_c must be dimensionally consistent.

Table 5.8 lists typical adhesive force of particles on various surfaces at different humidities. Note that these are nondissolving substances so

Table 5.8 Typical Adhesive Force of Particles

Substrate	Particle			Adhesive Force, dynes at Air RH		
	Material	Diameter (μm)	Type[a]	50-60%	90%	Removal
Glass	Glass	50	I	0.37	1.83	98%
Glass	Sand	50	I	0.76	2.06	98%
Glass	Coal	50	I	0.55	0.94	98%
Plexiglass	Glass	50	I	1.44	1.97	98%
Teflon	Sand	50	I	0.65	1.28	98%
Steel	Glass	40-60	I	1.64	–	98%
Steel	Glass	40-60	I	2.13×10^{-4}	–	50%
Steel	Glass	40-60	L	$21.7/cm^2$	–	–

[a]I = individual particles; L = powder layer.

humidity adds only water surface tension force. Adhesive force increases for dissimilar materials, increases with humidity and decreases with groups of particles. Overcoming adhesive force of particles on fabric filters is extremely significant as most particles must be removed to continue operating. Usually the dust is removed as agglomerates of substantial aggregate size. However, as cleaning energy is increased more is removed but aggregate size decreases.

5.7.2 Coalescence

The words agglomerate, coagulate and coalesce are usually used synonymously, which is accepted but incorrect as defined. Agglomeration relates more to formation of a ball by smaller solids while coagulation means the curdling or clotting of a liquid. Coalescing is used here to include all of these as a growing or merging together.

Particles in the atmosphere merge together because of Brownian motion, and the resulting collisions produce a decay rate equation:

$$-\frac{dN}{dt} = K N^2 \tag{5.42}$$

where N is particle number concentration per cm^3, t is time in seconds and K is a rate constant. When N_p is the initial particle concentration this gives

$$\frac{1}{N} - \frac{1}{N_p} = K t \tag{5.43}$$

Experimental studies give the value of K as about 0.51×10^{-10} cm^3/sec for typical particulate clouds and as high as 3.7×10^{-10} cm^3/sec for very fine material.

These equations do not account for particle size, which certainly influences diffusion and collision rate of confined particles. Haberl and Fusco[29] show rates of coalescence for various particle sizes and concentrations for monodisperse particles in a laminar system (Figure 5.14). This is an important relationship as the coalescence of small particles confined for even a relatively short period will significantly decrease the number concentration.

Particles in a turbulent system and particles with charges will coalesce much more quickly. Turbulent systems of a specific size, however, would have a smaller hold-up time. Removal of charged particles can be determined by electrostatic collection. Particles will unite by kinematic coagulation, which is the result of different size particles attaining different relative collection velocities in a collection device. This includes the conventional devices that have been discussed plus coupling with ultrasonic vibrations.

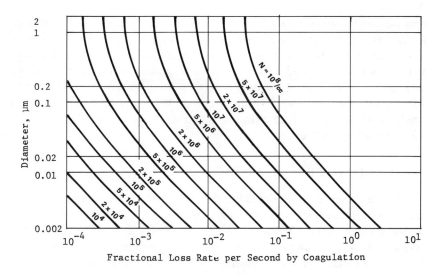

Figure 5.14 Fractional loss rate of monodisperse aerosols by coalescence.[29]

5.7.3 Evaporation and Condensation

Volatile liquids can nucleate, form drops and grow in size in super-saturated vapors or reduce in size and disappear by evaporation in un-saturated gases. Droplet diameter can be estimated after some time, t, knowing the initial diameter, d_i, using Frossling's modification of Maxwell's Equation:

$$d^2 - d_i{}^2 = \frac{8\,D_{AB}\,M\,t}{\rho_l\,R\,T}\,(p_\infty - p_0)(1 + 0.276\,Re_p{}^{1/2}\,Sc^{1/3}) \qquad (5.44)$$

where D_{AB} = vapor diffusivity, cm^2/sec

$\quad\quad\quad$ R = universal gas constant

$\quad\quad\quad$ p_∞ = vapor pressure in bulk of gas

$\quad\quad\quad$ p_0 = vapor pressure at droplet surface

This equation works best on large droplets (*e.g.*, ~ 1 mm). Adequate data are not available to describe the rate of change for fine particles, which can be very significant.

5.7.4 Adsorption

Adsorption is the taking up of a substance on the surface of a solid or liquid. These molecules are held to the surface by an inbalance in the electrical or chemical valence forces arising from interactions with surface or near surface molecules. Quantity of adsorbed vapor can be used to estimate surface area and therefore average particle size under controlled conditions. The presence of adsorbed molecules can modify the particle surface properties and affect any or all electrical charge capability, adhesion and evaporation rate. It can also result in the particle being an odorous pollutant as well as particulate matter. The presence of these adsorbed molecules can result in abnormal particle behavior and should be considered, especially when evaluating fine particles.

5.8 CUT DIAMETER

Cut diameter, d_{50}, as already defined, is the particle diameter for which collection efficiency (and penetration) equals 50%. This single diameter is useful for comparing efficiencies of various devices and of the same device under various operating conditions even when penetration does not reach 50%. To explain this, the following three examples are used following the procedure of Calvert et al.[30] and using their curve for venturi scrubbers, which is given as Figure 5.15. Data for these examples are summarized in Table 5.9.

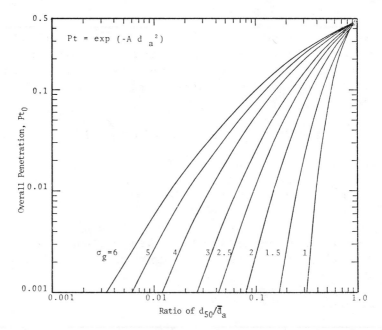

Figure 5.15 Venturi-type scrubber overall penetration versus the ratio aerodynamic cut diameter to aerodynamic mass mean diameter as a function of σ_g for log-normally distributed particles.[30]

Table 5.9 Summary of Data in Section 5.8 Example Problems

| | Example Comparisons | | | | | |
| | No. 1 | | No. 2 | | No. 3 | |
Data	Test 1	Test 2	Test 1	Test 2	Test 1	Test 2
σ_g	2.0	2.0	2.0	2.0	1.5	2.5
$\overline{d}_a$, μm	1.0	1.0	1.0	2.0	1.2	0.8
E_o	0.90	0.99	0.90	0.90	0.99	0.95
Pt_o	0.10	0.01	0.10	0.10	0.01	0.05
$d_{50}/\overline{d}_a$	0.34	0.17	0.34	0.34	0.25	0.14
d_{50}	0.34	0.17	0.34	0.68	0.30	0.11

1. Two venturi collection efficiency tests are made on a single dust with an aerodynamic mass mean diameter, $\overline{d}_a$, of 1 μm and a standard geometric deviation, σ_g, of 2.0. The first test gives an overall fractional penetration, Pt_o, of 0.1 (therefore overall efficiency, E_o, is 0.90) and the second test gives Pt_o of 0.01 (E_o = 0.99). It is obvious that the second test is more efficient, and this is also shown by calculating the aerodynamic cut diameter, d_{50}. Using Figure 5.15, the ratio of theoretical aerodynamic to aerodynamic mass mean diameter is about 0.34 for the first test and 0.17 for the second. Respective d_{50}'s are 0.34 μm and 0.17 μm for this dust showing the second test gives higher collection efficiencies.

2. Two tests are made on different venturi scrubbers. Test dust number one has σ_g = 2.0, $\overline{d}_a$ = 1 μm and scrubber Pt_o = 0.1. The second test has σ_g = 2.0, $\overline{d}_a$ = 2 μm and Pt_o = 0.1. It is obvious from inspection of the data that test one is more efficient. Using the same procedure as in example 1, $d_{50}/\overline{d}_a \cong$ 0.34 for both, but d_{50} is 0.34 μm for test one and 0.68 μm for test two showing number one is more efficient.

3. Two tests are made on different venturis with different dusts as shown in Table 5.9. The first test may appear more efficient but the dusts are different and no direct comparison can be made using efficiencies. Determination of d_{50} shows that actually the second system could be more efficient if run on the same dust.

Collection efficiency data from improved collection systems shows that the use of some "equivalent" cut diameter must be obtained to rate them because overall penetration does not always reach 0.50. The equivalent cut diameter suggested[30] is defined as the theoretical cut diameter of a venturi-type scrubber required to give the same overall penetration for the particular particle size distribution under consideration. It is found using the observed Pt_O for the test unit and the dust $\bar{d}_a$ and σ_g and Figure 5.15, as in the examples above.

REFERENCES

1. Billings, C. E., *et al.* "Handbook of Fabric Filter Technology Vol. I, Fabric Filter Systems Study," National Technical Information Service No. PB 200 648 (December 1970).
2. McIlvaine, R. W. (Ed.) *The Fabric Filter Manual* (The McIlvaine Company, 1976), Chapter III.
3. Williams, C. E., T. Hatch and L. Greenberg. "Determination of Cloth Area for Industrial Air Filters," *Heating, Piping & Air Conditioning* 12:259-63 (1940).
4. Billings, C. E. and J. E. Wilder. "Major Applications of Fabric Filters and Associated Problems," *Proceedings U.S. EPA Symposium on Control of Fine-Particulate Emission from Industrial Sources,* San Francisco (January 1974).
5. Nichols, G. B. *The Electrostatic Precipitator Manual* (The McIlvaine Company, 1976), Chapter II.
6. McLean, K. J. "Factors Affecting the Resistivity of a Particulate Layer in Electrostatic Precipitators," *J. Air Poll. Control Assoc.* 26(9):866-870 (1976).
7. Sproull, W. T. "Fundamentals of Electrode Rapping in Industrial Electrical Precipitators," *J. Air Poll. Control Assoc.* 15:50-55 (1965).
8. Calbert, S. *et al.* *Scrubber Handbook* (APT, Inc., 1972).
9. Oglesby, S., Jr. and G. B. Nichols. "A Manual of Electrostatic Precipitator Technology, Part I, Fundamentals and Part II, Application Areas," Southern Research Institute (August 1970).
10. Cross, F. L., Jr. and H. E. Hesketh. *Handbook for the Operation and Maintenance of Air Pollution Control Equipment* (Technomic Publishing Company, Inc., 1975).
11. Golovin, M. N. and A. A. Putnam. *Ind. Eng. Chem. Fund.* 1:264 (1962).
12. Perry, R. H., C. H. Chilton and S. O. Kirkpatrick. *Chemical Engineer's Handbook* (New York: McGraw-Hill Book Co., 1963).
13. Hesketh, H. E. "Fine Particle Collection Efficiency Related to Pressure Drop, Scrubbant and Particle Properties, and Contact Mechanism," *J. Air Poll. Control Assoc.* 24(10):939 (1974).
14. Cooper, D. W., L. W. Parker and E. Mallove. "Overview of EPA IERL-RTP Scrubber Programs," EPA-600/2-75-054 (September 1975).
15. Walton, W. H. and A. Woodrock. "The Suppression of Airborne Dust by Water Spray," *Internat. J. Air Poll.* 3:129-153 (1960).
16. Melcher, J. R. and K. S. Sachar. "Charged Droplet Scrubbing of Submicron Particulate," EPA-650/2-74-075 (August 1974).
17. Abbott, J. H. and D. C. Drehmel. "Control of Fine Particulate Emissions," *Chem. Eng. Prog.* 72(12):47 (1976).
18. Leith, D. and M. E. First. "Filter Cake Redeposition in a Pulse-Jet Fabric Filter," paper 76-31.4, 69th meeting of the Air Pollution Control Association, Portland, Oregon, June, 1976.

19. Billings, C. E., M. E. First, R. Dennis and L. Silverman. "Laboratory Performance of Fabric Dust and Fume Collectors," U.S. Atomic Energy Commission Report No. NYO-1590-R (August 1954, revised January 1961).
20. Gooch, J. P., J. R. McDonald and S. Oglesby, Jr. "A Mathematical Model of Electrostatic Precipitation," U.S. Environmental Protection Agency, Publication No. EPA 650/2-75-016.
21. Bell, C. G. and W. Strauss. "Effectiveness of Vertical Mist Eliminators in a Cross Flow Scrubber," *J. Air Poll. Control Assoc.* 23(11):967 (1973).
22. Calvert, S., S. Yung and J. J. Leung. "Entrainment Separators for Scrubbers," Initial Report, EPA-650/2-74-119b; and S. Calvert, I. Jashnani, S. Yung and S. Stalberg. Initial Report, EPA-650/2-74-119a, U.S. Environmental Protection Agency, Office of Research and Development (1974).
23. York, O. H. "Performance of Wire Mesh Demisters," *Chem. Eng. Prog.* 50(8):421 (1954).
24. Hougen, O. A., K. M. Watson and R. A. Ragatz. *Chemical Process Principles, Part II, Thermodynamics* (New York: John Wiley & Sons, Inc., 1965).
25. Smith, J. M. and H. C. Van Ness. *Introduction to Chemical Engineering Thermodynamics* 2nd ed. (New York: McGraw-Hill Book Co., 1959).
26. Bird, R. B., W. E. Stewart and E. N. Lightfoot. *Transport Phenomena* (New York: John Wiley & Sons, Inc., 1960).
27. Fuchs, N. A. *The Mechanics of Aerosols* (London: Pergamon Press, 1964).
28. Krupp, H. *Advances in Colloid and Interface Science (Particle Adhesion Theory and Experiment)*(New York: John Wiley & Sons, Inc., 1967).
29. Haberl, J. B. and S. J. Fusco. "Condensation Nuclei Counters: Theory and Principles of Operation," General Electric Technical Information Series, No. 70-POD 12 (1970).
30. Calvert, S. *et al.* "APS Electrostatic Scrubber Evaluation," EPA-600/2-76-154a (June 1976).

CHAPTER 6

SIZE MEASUREMENT

6.1 GENERAL

Behavior of fine particles in gaseous media can be difficult to predict because of the numerous complex factors mentioned previously. Obtaining accurate size, shape, index of refraction, density and aerodynamic measurements on these particles in their actual dynamic state is often a time-consuming and costly task. In addition to the particle physical and behavioral variations, a wide variety of testing conditions are likely to be encountered and must be accounted for. This includes gas temperature, pressure and composition including moisture content. Added complexities exist that have not even been discussed, such as variations in particulate concentrations by orders of magnitude from inlet to outlet of a collection device. This plus the useful range limitations of test instruments may make ít necessary for several devices to be used to obtain complete information on the particle properties.

Particle sizing measurements to provide data on quantity of emissions and for determination of collection device fractional collection efficiencies are presented in this chapter. Although numerous methods are available and being developed, the currently practical particle sizing techniques fall into three categories: inertial, diffusional and optical. Some of these procedures can size particles $\geq 0.002 \ \mu m$. There are limitations and advantages with each procedure and these are discussed in the next several sections. Briefly, however, impactors are prone to overloading and reentrainment, and emission concentrations are frequently too high for direct measurement by optical and diffusional counters.

6.1.1 Particle Sampling

Obtaining valid particle samples is not always easy. In addition to the proper sizing device, certain procedures and experience are necessary to

151

obtain the needed gas and particulate information. Many of the procedures are local and/or federal requirements. The federal Environmental Protection Agency regulations are published and updated periodically in the *Federal Register* (Title 40, Code of Federal Regulations, parts 53, 60, etc.) and in the form of manuals.

Isokinetic sampling of a moving gas stream is usually one requirement for trying to obtain the most accurate particulate sample. This means that gas in the system is moving past the sampling probe at the same velocity as the gas sample entering the mouth of the probe. Lower sampling velocities result in impaction of larger particles into the probe opening with a subsequent skewing of the size distribution data. Higher sampling velocities can result in excess smaller particles.

Theoretically it is not possible to obtain an exact particulate sample distribution even by isokinetic sampling, because of effects such as gravitational sedimentation; surface drag; presence of an impaction surface even though probe ends are tapered, sharp and smooth; particle coalescence and evaporation; gradients of concentration, temperature and pressure; and other effects.

Neglecting effects other than those due to anisokinetic sampling, particle size, probe size, and gravity, and assuming the sample probe is absolutely parallel to the gas flow, Ruping[2] developed a relationship in the German Standards for isokinetic sampling and system parameters. This is given as Figure 6.1 and shows curves of 1% and 5% sampling error as a function of velocity ratio, v_∞/v_s, and dimensionless parameter, k:

$$k = \frac{g\,D}{2\,v_\infty\,v_s} \tag{6.1}$$

where D is probe inside diameter, v_∞ is approach gas velocity and v_s is particle sampling velocity. To obtain an error of $< 5\%$, velocity must be within about 6% of isokinetic and for $<1\%$ error velocity must be within 2%.

Oglesby[3] points out that although it is not possible to aim the sample probe correctly for all particle sizes, an error of $5°$ or less is normally not significant. At very low gas velocities, isokinetic sampling and probe angle deviations can cause disastrously large errors.

6.1.2 Gas Sampling

Particle emission rate and collection efficiency evaluations require that accurate gas measurements be obtained. The most common technique for obtaining accurate gas velocities is use of a standard pitot. A good commercial pitot can provide gas flow data for impact pressures from about 0.025-25 cm water gauge. This assumes that the device is properly operated, clean and in reasonably good condition, and that accurate pressure read-out

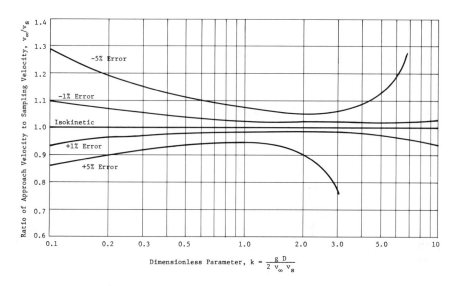

Figure 6.1 Isokinetic sampling accuracy.

capacity is used. For air at standard conditions this corresponds to flow velocities from about 200-10,000 cm/sec. Figure 6.2 gives values for air flow over some of this range.

Fortunately, pitot tubes can give accurate gas flow measurements even though dust concentrations are extremely high. Hoffman[4] showed that no measurable difference is noted at concentrations of quartz dust in air up to 1 kg/m³. This is true only if the pitot and lines are clean of deposits and liquid, the manometer is accurately calibrated and the manometer liquid is not fluctuating too rapidly.

Gas flow velocities, v, at any point can be measured from pitot impact pressure using

$$v = \frac{T}{P}\sqrt{\frac{2R\,P_S\,\Delta P}{M\,T_S}} = 4974\sqrt{\frac{P_S\,\Delta P}{M\,T_S}} \qquad (6.2)$$

where v = cm/sec
T = 293.16 °K
P = 760 torr = 760 mm Hg @ 0°C
R = universal gas constant, 83.14 x 10⁶ g cm²/(g-mole-sec² °K)
P_S = static pressure, torr
ΔP = pitot differential pressure, torr
T_S = static temperature, °K
M = gas molecular weight (wet), g/g-mole

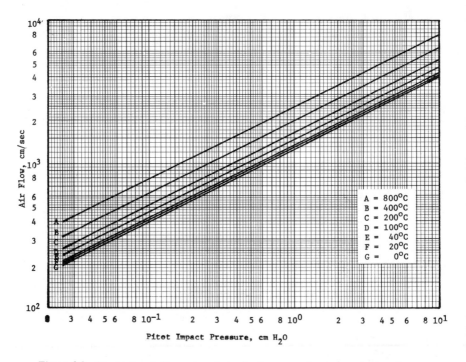

Figure 6.2 Pitot tube air flow rates at standard pressure for various temperatures.

The molecular weight of the gas must be determined from the gas composition. This can be done using the chain rule, assuming ideal gases. For example, average molecular weight of wet flue gas, M_{FG}, is

$$M_{FG} = (1 - Y_{H_2O}) [32 Y_{O_2} + 44.01 Y_{CO_2} + 0.04 + 28.01(0.99 - Y_{O_2} - Y_{CO_2})]$$

$$+ 18.01 Y_{H_2O} \qquad (6.3)$$

where Y stands for mole fraction in the gas phase, and oxygen (O_2) and carbon dioxide (CO_2) are *dry* gas values obtained for example by Orsat analysis. The term $(0.99 - Y_{O_2} - Y_{CO_2})$ in this equation represents mole fraction nitrogen. It is assumed that no measurable CO is present.

6.1.3 Errors and Calculation

In obtaining both the particulate and gas samples, three categories of errors can occur and spoil the best data. These are mistakes, systematic errors and random errors. Mistakes are usually more obvious and hopefully will be corrected during a recheck of the work. Systematic errors are reproducible errors such as a shift in values because of change in instrument voltage, repeated use of an incorrect conversion factor or an incorrectly calibrated meter. Random errors are inconsistent and may be related to random perturbations of the system under test or the testing system. These errors usually cannot be corrected, making it necessary for a number of tests to be run to determine the needed information from the distribution of data.

One method for determining whether a significant difference exists in data is by the use of Student's "t" test for small samples. As an example, an arithmetic mean particle emission, $\overline{E}$, can be obtained from a given series of test data where the number of runs or individual data points, n, is > 3:

$$\overline{E} = \frac{\sum_{i=1}^{n} E_i}{n} \qquad (6.4)$$

where E_i is the emission rate from each run. The sample variance, S^2, is then

$$S^2 = \frac{\sum_{i=1}^{n} (E_i - \overline{E})^2}{n - 1} \qquad (6.5)$$

Values for $\overline{E}$ and S^2 obtained from two different conditions (*e.g.*, before, a, and after, b, some change) can be compared to see if there is any difference in emissions as measured using the pooled estimate, S_p, and the test statistic, t:

$$S_p = \left[\frac{(n_a - 1)(S_a^2) + (n_b - 1)(S_b^2)}{n_a + n_b - 2} \right]^{1/2} \qquad (6.6)$$

$$t = \frac{\overline{E}_b - \overline{E}_a}{S_p \left[\frac{1}{n_a} + \frac{1}{n_b} \right]^{1/2}} \qquad (6.7)$$

In this comparison, if $\overline{E}_b > \overline{E}_a$ *and* $t > t'$ then with 95% confidence there is a significant difference. An increase in emissions has occurred in this example and not simply the presence of random errors. If both conditions are not met, random errors account for the differences. Several values of t' are given for the 95% confidence level in Table 6.1. Degrees of freedom equal $(n_a + n_b - 2)$. See standard tables for other confidence levels and for more than 8 degrees of freedom.

Knowing the data are valid, it is sometimes useful to know whether certain factors are related. For example, is scrubbing liquid temperature related to collection efficiency in a particular device? This can be determined by use of a correlation coefficient, R. Linear correlation is the

Table 6.1 Student's "t" Test Comparison Statistics for 95% Confidence Level

No. of Degrees of Freedom	t'
2	2.920
3	2.353
4	2.132
5	2.015
6	1.943
7	1.895
8	1.800

easiest to describe and is explained here. Other types of correlation can be determined using procedures such as those described by Meyer.[5]

The linear correlation coefficient is found by either of the following equations:

$$R = \frac{(1/n) \sum_{i=1}^{n} (x_i - \bar{x})(y_i - \bar{y})}{[(1/n) \sum (x_i - \bar{x})^2]^{1/2} [(1/n) \sum (y_i - \bar{y})^2]^{1/2}} \tag{6.8}$$

$$R = \frac{n \sum x_i y_i - \sum x_i \sum y_i}{[n \sum x_i^2 - (\sum x_i)^2]^{1/2} [n \sum y_i^2 - (\sum y_i)^2]^{1/2}} \tag{6.9}$$

where n is the number of individual points (x_i, y_i).

The numerical value of R shows whether two quantities X and Y have a linear correlation and the degree of correlation. A value close to ± 1 shows nearly perfect correlation (direct for +1 and indirect for –1). Values close to zero show they are uncorrelated. As an example, specific gravity and melting point of metals have an R of 0.864 showing these properties are correlated.

6.2 CASCADE IMPACTORS

6.2.1 Arrangements

Inertial devices including impactors, impingers, cyclones and centrifuges are used for particle sampling and size estimation. The cascade impactor, because of its compact arrangement and ruggedness has generally been accepted as the most suitable inertial device for obtaining mass and mass distribution measurements. A survey of the literature[6-13] gives an abbreviated history of the development of the various cascade impactors showing that a variety of design configurations can be used. These range from an impactor with long (4.763 cm) narrow, rectangular throat jets and rotating 3.8-cm diameter cylindrical collection surfaces to the more common orifice jets with flat tray collection surface arrangements, all inside a cylindrical body about 7.4 cm in diameter by 25 cm long overall including sampling nozzle as shown in Figure 6.3. This figure is a typical Mark III-type cascade impactor and is used in the following discussion as the example.

Details and dimensions of the University of Washington Mark III cascade impactor are given in Table 6.2. Jet depth refers to the thickness of the plate from which the jets are made (*e.g.,* by machining, abrasive drilling, laser

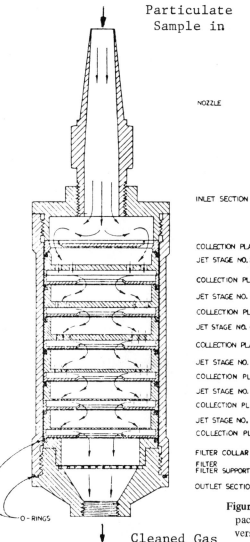

Particulate
Sample in

NOZZLE

INLET SECTION

COLLECTION PLATE NO. 1
JET STAGE NO. 2

COLLECTION PLATE NO. 2
JET STAGE NO. 3

COLLECTION PLATE NO. 3
JET STAGE NO. 4

COLLECTION PLATE NO. 4
JET STAGE NO. 5

COLLECTION PLATE NO. 5
JET STAGE NO. 6

COLLECTION PLATE NO. 6
JET STAGE NO. 7
COLLECTION PLATE NO. 7

FILTER COLLAR
FILTER
FILTER SUPPORT PLATE

OUTLET SECTION

O - RINGS

Cleaned Gas
to Meters

Figure 6.3 Cascade impactor (Mark III, University of Washington type).

Table 6.2 University of Washington Mark III Cascade Impactor Details and Dimensions

Stage	Number of Jets	Jet Diameter (mm)	Jet Depth (mm)	Jet-to-Plate Clearance (mm)	Jet Depth to Jet Diameter Ratio	Jet-to-Plate to Jet Diameter Ratio
1	1	18.237	38.1	14.2	2.08	0.78
2	6	5.791	3.175	6.477	0.548	1.12
3	12	2.438	3.175	3.175	1.30	1.30
4	90	0.787	3.175	3.175	4.03	4.03
5	110	0.508	1.600	3.175	3.15	6.25
6	110	0.343	0.762	3.175	2.22	9.26
7	90	0.254	0.762	3.175	3.00	12.50

beam, etc.). Jet-to-plate clearance is distance from bottom of jet to top of collection plate. In actual use, this could be reduced by the thickness of the insertion plates (~ 0.13 mm) and grease layer when they are used.

Referring to Figure 6.3, the sample enters through the interchangeable nozzle, which must be appropriately chosen by the user for isokinetic sampling. The nozzle itself serves as the first-stage jet. Large particles impact from the relatively low-velocity gases on the first collection plate while the smaller particles move with the gases through an annulus to the second stage. This stage has a number of smaller jets arranged to increase the gas velocity, causing smaller particles to be captured on the second collection plate. The process continues with smaller holes, higher velocities and smaller particles captured on each succeeding stage so as to aerodynamically fractionate the particulate sample. In this device the collection plates, except for #1, are donut-shaped so the gases escape to the next stage through the center. A final filter is used after the last stage to remove all residual particulates. This filter can be of paper, glass, porous metal foil, porous organics such as plastics and Teflon or other suitable material.

6.2.2 Operation

The particles are collected on preweighed pans or preweighed inserts placed on the pans. The device can be operated in any position because the impacted particles adhere to the collection surface. To help insure this, a thin layer of grease (< 30 mg per stage) can be added to the surface before dessication and weighing. It may be worthwhile to bake these surfaces with the grease for several hours before weighing. The grease must be evenly distributed so it will be only "fingerprint" thick or it will be blown off, creating errors on several stages. Smith *et al.*[14] report that grease blow-off can occur at velocities > 65 m/sec. The grease must also be nonvolatile and nonreactive with the system. Dow Corning high–vacuum grease is sometimes used at temperatures $< 250°C$.

Excessive velocities at the collection surface and/or excessive accumulation of particles can result in a blowing off of deposited sample. Smith reports that this becomes severe at velocities greater than about 35 m/sec and can occur at lower velocities for some materials. This can be minimized by operating within the specified flow rates of a well-designed device and by accumulating no more than about 10 mg of sample per plate. Sampling times must be estimated based on flow rates and projected concentrations to provide enough but not too much sample. A comparison of maximum velocities to prevent reentrainment of a test particulate in a Mark III impactor are given in Table 6.3 for various nozzle sizes. This amounts to about 1.3 m³/hr (0.8 acfm) with grease and 0.5 m³/hr (0.3 acfm) without grease and applies to stage one only.

Table 6.3   Maximum Nozzle Velocity in Mark III Impactor to Prevent Reentrainment of Test Particulate[14]

Nozzle Inside Diameter (mm)	Maximum Gas Velocity (m/sec)	
	With Greased Surface	No Surface Grease
12.7	3.05	1.13
9.5	5.49	2.04
9	6.10	2.32
8	7.63	2.87
7	10.1	3.97
6.4	12.2	4.58
6	13.4	5.35
5	20.4	7.63

Collection pans weigh 40-60 g each, and the stainless steel pan cover inserts weigh 0.6-0.9 g each. It becomes obvious with sample mass per stage of only 0.5-10 mg and a grease mass of up to 30 mg that very careful handling, preparation, drying and weighing are necessary, even if losses do not occur elsewhere. A balance with a sensitivity of 0.1 mg is useful for this work.

Moisture droplets will be fractionated just as any particulate; however, special care is needed to see that they do not run off the plates, evaporate or become lost in some other manner. Ideally, an impactor with a deep pan is required for sampling liquid particles. Condensation inside the impactor cannot be tolerated so the device must be preheated externally or closed off and placed in the hot test system for about 30 minutes prior to starting the test to permit the entire unit to warm to the operating temperature or at least above the dew point.

6.2.3 Design

A detailed analysis of multi-jet cascade impactors made by Cohen and Montan[10] produced the following recommended design parameters:

1. Jet Re_f should be $> 100/$(jet depth to jet diameter ratio);
2. Jet Re_f should be < 3200 to remain in Stokes region;
3. Jet velocity should be > 10 times terminal settling velocity of d_{50} particles;
4. Jet velocity should be $< 1.1 \times 10^4$ cm/sec for incompressible flow;
5. Optimum jet-to-plate to jet diameter ratio is in the 1-3 range; and
6. Jet depth to jet diameter should be $\geqslant 1$.

The expression d_{50} is the aerodynamic cut diameter and indicates the size of particle captured on that stage with a 50% collection efficiency.

Mercer and Stafford[15] developed an expression, β, to show divergence of the gas stream leaving a jet and defined it as

$$\beta = \frac{2 s}{D_j} \tag{6.10}$$

where s is bottom of jet to top of collection plate distance and D_j is diameter of jet. Calvert et al.[16] showed that β is related to $\sqrt{K_{50}}$, where K_{50} is the impaction parameter corresponding to 50% collection efficiency. Impaction efficiency curves as a function of $\sqrt{K_I}$ are steeper for round jets than for rectangular jets and the effect of β is less significant. For round jets, $\sqrt{K_{50}}$ increases as β increases for $\beta < 10$. For rectangular jets this is true for $\beta < 4$. Above these values of β, $\sqrt{K_{50}}$ is a constant.

Theory and experimental data show that for a given jet size and arrangement, β is fixed and if < 10 for round jets and < 4 for rectangular jets there is normally only one K_{50}, regardless of particle size and gas velocity. From the definition of impaction parameter, this gives

$$\sqrt{K_{50}} = \left(\frac{\rho_p v_j C}{9 \mu_g D_j} \right)^{\frac{1}{2}} d_{50} \tag{6.11}$$

where v_j is the jet velocity. When the jet velocity is less than about one-third the Mach number, flow is incompressible and velocity equals volumetric flow rate into the jet divided by jet cross-sectional area:

$$v_j = \frac{4 Q}{\pi D_j{}^2} \tag{6.12}$$

If $v_j > 1/3$ Ma, then Equation (6.12) must be multiplied by the ratio of jet pressure in/pressure out.

For incompressible flow and for particles in the Cunningham-Stokes Regimes, the Cunningham factor, C, is given by equations in Section 1.6. For compressible flow, the jet pressure drop can be estimated using

$$\Delta P = \frac{\rho_g v_j^2}{2} C_4 \qquad (6.13)$$

where C_4 is a dimensionless flow constant which may equal 1 for some stages but can be as much as 20.

Equation (6.11) can be solved to obtain d_{50} for a given stage using the relationships given above and using a value of $\sqrt{K_{50}}$. Based on available data, an average $\sqrt{K_{50}}$ of about 0.44 could be assumed for round jets and about 0.80 for rectangular jets. These values can be determined experimentally for a given jet configuration using monodisperse spherical particles with a density of 1.0, so diameters are aerodynamic diameters.

6.2.4 Efficiencies

Calvert[17] lists ratios of jet-to-plate to jet diameter, s/D_j, for the smaller four stages of several cascade impactors (Table 6.4). Averaged results of collection efficiencies for these four stages as measured by Calvert are then compared to published theoretical and experimental data in Figure 6.4. The effect of the ratio s/D_j becomes apparent when curves C and D are compared. The actual efficiencies measured by Calvert for a University of Washington Mark III impactor are given as Figure 6.5 and for comparison, the Mark III calibration curve given by Pilat[18] for a seven-stage impactor is shown in Figure 6.6.

Table 6.4 Jet-to-Plate to Jet Diameter Ratios for Cascade Impactors[17]

Impactor	Stage			
	4	5	6	7
A. P. T., M - I	3.1	2.4	3.5	4.6
University of Washington Mark III	4.0	6.2	9.2	12.5
Anderson, nonviable	4.7	7.3	9.8	9.8

Remember that all diameters noted here are aerodynamic diameters and real particle size, if required, must be calculated from this. Raabe[19] proposed use of the "Lovelace" aerodynamic resistance diameter, d_{ar}. This is named because of the common use at the Lovelace Foundation for Inhalation Research. It is a viscous resistance diameter that describes respirable

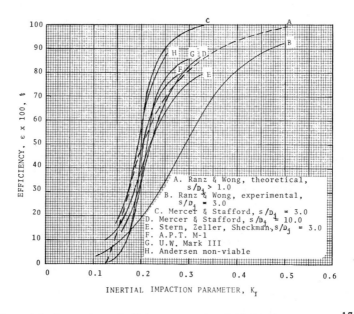

Figure 6.4 Comparison of efficiency versus inertial impaction parameter.[17]

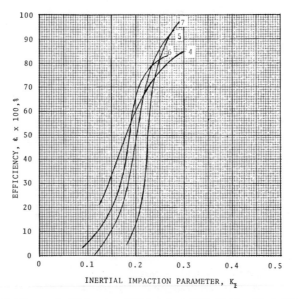

Figure 6.5 Measured efficiencies of University of Washington Mark III impactor (stages 4-7) versus impaction parameter.[17]

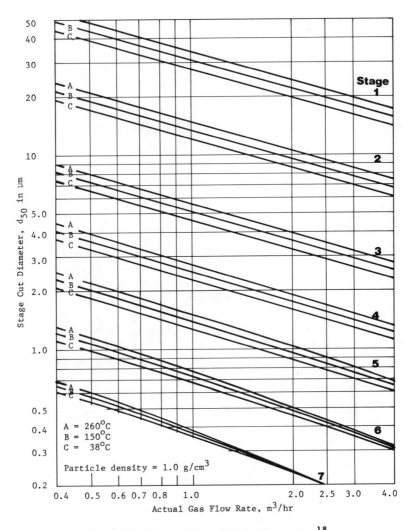

Figure 6.6 Calculated d_{50} of Mark III impactor.[18]

particles, which are those in the Stokes and Cunningham Regimes and have $Re_p < 0.5$ and geometric diameter $> 0.3 \mu m$. It is defined by

$$d_{ar} = d \sqrt{\rho_p dC} / \sqrt{1 \text{ g/cm}^3} \qquad (6.14)$$

which makes the units rather unorthodox. Considering that gas viscosity is nearly independent of pressure and depends on temperature, d_{ar} varies only with temperature. Calibration of an inertial sampler at temperature T_1 gives d_{ar1} and at T_2 the new calibration would be

$$d_{ar2} = d_{ar1} \left(\frac{\mu_{g2}}{\mu_{g1}} \right)^{\frac{1}{2}} \tag{6.15}$$

By contrast, aerodynamic diameter changes with both temperature and pressure.

Measurement errors can develop because of problems such as those noted in this and the previous section. In addition, significant errors in impactors have been measured due to weight gain or loss by reactions of impactor surfaces, grease, and filters with the gases; by deposits within the body of the impactor; and by pluggage or erosion of the jets.[14,20,21]

6.2.5 Data Reduction

Impactor raw data must be converted to usable form to be of value. Sampling conditions, measured flow rates and mass collected over the specific sampling period must be known, and if average values are acceptable, particle size and size distribution can be determined. Actual gas flow rate through the sampler, Q_a, is determined from measured flow corrected for moisture loss, temperature, pressure and sampling time. Impactor stage cut diameters (size collected at 50% efficiency) are determined using measured or calculated efficiency calibration curves such as Figure 6.6 or by calculation, for example, where air at standard conditions is assumed:

$$d_{50} = 2.72 \times 10^{-3} \left[\frac{\mu_g \, D_j^3 \, P_s \, N_j}{\rho_p \, Q_a \, P_o \, C} \right]^{\frac{1}{2}} C_5 \tag{6.16}$$

where
D_j = jet diameter, mm
P_s = static pressure immediately upstream of impactor stage, absolute
N_j = number of jets at that stage
Q_a = actual flow rate at that stage, m^3/hr
P_o = pressure at impactor inlet, absolute
C_5 = empirical const. dependent on impactor and stage, use $C_5 = 1$ if data not available

This is an iterative trial-and-error solution as C must be known to obtain d_{50} and C is a function of d_{50}.

Particle size distribution can be represented by either a differential or cumulative basic. The differential distribution assumes that all mass captured on an impaction stage consists of material having aerodynamic diameters $> d_{50}$ of that stage and $< d_{50}$ of the preceding stage. The intervals between stage d_{50}'s are logarithmically related so usually a semi-log or log-log plot is made of $(d \, M)/(d \log d)$ versus $\log d_{geo}$. The change in mass per stage, dM, is normalized to a value of mass per unit volume of dry gas using the concentration per unit volume of dry gas sampled, C_M. This gives for the stages i and i + 1

$$\frac{dM}{d \log d} = \frac{(\epsilon_i + 1)C_M}{\log (d_{50})_i - \log (d_{50})_{i+1}} \qquad (6.17)$$

where ϵ_i is the mass fraction collected on stage i.

The geometric mean diameter, d_{geo}, is

$$d_{geo} = \sqrt{(d_{50})_i \, (d_{50})_{i+1}} \qquad (6.18)$$

This method of plotting is convenient for obtaining fractional collection efficiency for some pollution control device when inlet and outlet impactor samples are taken. A comparison of inlet and outlet plots shows the material in the gas sampled at any given size. The ratio of outlet concentration over inlet concentration at a specific size is penetration or $(1 - \epsilon_i)$ of the control device for that size particle.

Cumulative particle size distributions are obtained by summing the mass collected on each stage and the filter on a cumulative basis and plotting the mass below a given size on log-probability plots, as discussed in Chapter 1. The plot could be obtained by integration of the previous method, but a more rigorous procedure is to attempt to account for all factors. This considers the fact that a single-size particle can be collected on several stages and uses calculated "actual" stage loadings which are a function of initial concentration and particles captured on preceding stages. A method developed by Picknett[22] has been programmed for computer determination of "best fit" cumulative distribution. It must account for the type of impactor and operating characteristics.

A simplified technique for obtaining immediate estimates of cumulative particle size distribution that agrees closely with some computerized procedures is suggested. This procedure assumes that half the material on the stage is larger than the d_{50}. This is not true, but it is a good approximation for fine (and especially submicron) particles. Half the mass percent on the last stage (above the filter) is added to the weight percent on the filter and this *value* is plotted against the d_{50} of the last stage. The remaining half of this mass plus half the mass caught on the next larger stage is added to this *value* and this new cumulative mass percent is now plotted versus d_{50} for this next larger stage. The process is repeated for all stages to obtain a complete estimate of cumulative particle size distribution.

6.2.6 Extending Usefulness

There are several ways to obtain additional data from impactors. Size of particles on the collection plates can be estimated mathematically as described. The size of solid particles on the plates and on porous metallic or organic final filters can be physically determined by microscopy or by electron microscope. Chemical composition of the collected material can be obtained to a limited degree by the use of a scanning electron microscope. The small sample size limits the use of wet chemistry analyses, although they can be made occasionally. In addition, a visual comparison of the samples on the various stages often shows the presence of a bimodal sample. Size distribution, visual observations, electron microscope and scanning electron microscope evaluations of impactor collected fly ash samples show that the larger fly ash in Figure 1.2 is unburned coal dust and the smaller is ash.[21]

Combination of impactors with other devices such as diffusion batteries and optical counters can extend the size distribution data to include particles smaller than the present impactor lower limit of about 0.2 μm. Parallel use of the final filter to obtain mass, and an optical counter in place of the final filter to obtain a number count is an example of this. Attempts are being made to develop low-pressure impactors for sizing to 0.02 μm, but they have not yet been successful.

6.3 DIFFUSION BATTERIES

6.3.1 Systems

Diffusion batteries can provide particle size distributions for particles with diameters in ranges of 0.002-0.2 μm. The system arrangements vary. Some consist of a series of screen stages while others contain a number of equally spaced long, narrow, parallel plates or a bundle of small-bore parallel tubes of equal diameter. Particles in the 0.002-0.2 μm range are called condensation nuclei with the smaller ones being called Aitken nuclei. Diffusivities range from 10^{-2} cm^2/sec for 0.002 μm to 2 x 10^{-6} cm^2/sec for 0.2 μm in air at standard conditions and for ρ_p of 1 g/cm^3.

Particles in these devices diffuse to the screen or wall surfaces, and it is assumed that they then adhere and are thus removed from the gas sample stream. Diffusivities decrease with increasing particle size, so extent of penetration in a battery depends on particle size. In a staged series, the first stage removes the smallest particles, and the larger particles are removed in the last stage.

The penetration from each stage is measured with a continuous-flow-type condensation nuclei counter. Variations both in number of stages or in length and number of channels or tubes and in gas flow rate are used as means of obtaining different particle penetration rates. This is necessary to prevent overloading the counter.

The TSI model 3040 is a new compact device and is used as an example diffusion battery. This 10-stage model is shown in Figure 6.7. The 8 cm x 24 cm-long device contains 55 stainless screens, 4 cm in diameter with 20 μm openings (635 mesh). It provides 10 size intervals for 0.002-0.1 μm particles at a flow rate of 4 ℓpm. In the device, particles in laminar flow diffuse to and adhere on the screens.

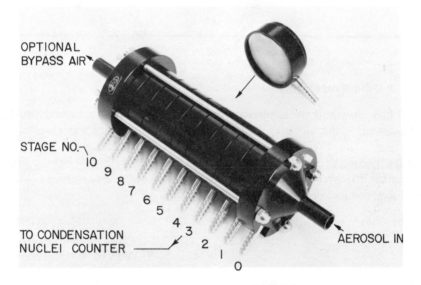

Figure 6.7 Ten-stage diffusion battery (Courtesy Thermosystems Inc.).

Table 6.5 notes what is obtained when a typical ambient aerosol with a number median diameter of about 0.04 μm is passed through a Model 3040 Diffusion Battery at 4 ℓpm. Particle diffusivity as a function of size is vividly shown with this aerosol. Nearly the entire smallest half is removed in the first three stages, which account for only about 11% of the total collection surface. These values are by number and not by mass as would be obtained in an impactor. While this screen–type device has the advantage of compactness, it cannot evaluate effect of sedimentation of larger particles. Plate or tube batteries turned on their sides would give this effect.

Table 6.5   Ambient Aerosol Removal in a Ten-Stage TSI Model 3040 Diffusion Battery

No. Screens per Stage	Cumulative No. Screens	Overall Percent Penetration by Number
1	1	90.0
2	3	74.0
3	6	55.0
4	10	42.0
5	15	30.0
6	21	22.0
7	28	16.0
8	36	11.8
9	45	8.8
10	55	6.6

6.3.2 Diffusional Sizing

Size distribution calculations for diffusional sizing were developed by Townsend[23] and Gormley and Kennedy.[24]   This theory was checked by Thomas[25] for tubular and parallel plate batteries and applied for screen series by Sinclair and Hoopes.[26]   Thomas has empirically developed a single equation to cover particle penetration over the entire size range of 0.002-0.2 μm for tubes or columnated hole structures:

$$Pt = 0.819 \exp(-3.65\alpha) + 0.097 \exp(-22.3\alpha) + 0.032 \exp(-57\alpha) + 0.027 \exp(-123\alpha)$$

$$+ 0.025 \exp(-750\alpha) \qquad (6.19)$$

where Pt =   fractional penetration
  α = $\pi D_{PM} \ell / Q$
  ℓ =   length of tube, cm
  D_{PM} =   particle diffusivity, cm^2/sec
  Q =   volumetric flow rate per tube, cm^3/sec

Application to the screen–stage diffusion batteries can be made by using a cumulative equivalent length because the diffusional collection mechanism of fibrous filters and wire screens with laminar flow is the same as for flow through round tubes.   This equivalent length is approximately the actual length, times, number of holes in each screen, times number of screens per stage; but it is better to fit cross plots of penetration versus number of screens and penetration versus tubular battery length to obtain an exact equivalent length.

Aerosol charge neutralizers are sometimes recommended for use with diffusion batteries to prevent electrostatic precipitation from influencing diffusional deposition efficiency. Sinclair *et al.*[27] report that the difference in diffusion penetration between a natural charged and a neutralized aerosol was insignificant except for aerosols smaller than about 0.01 μm.

6.4 CONDENSATION NUCLEI COUNTERS

Condensation nuclei counters can be used with other devices or alone. Counters follow diffusion batteries to provide the particle count data. They are also used alone or in parallel with other systems to give number concentration. The concentration limit for these systems is about 10^7 particles/cm^3 so it is often necessary to dilute the sample before it enters the counter. Problems of sample condensation and agglomeration can be severe so the dilution gas should be carefully cleaned and dried. Particles $\geqslant$ 0.002 μm are measured by condensation counters.

Condensation nuclei counters are extremely sensitive and can give very accurate results. They operate on the principle of a cloud chamber. Water obtained by supersaturating the air to about 300% condenses upon the submicron particles producing micron-sized droplets. The humidified air sample is diverted into the cloud chamber where a fixed volume expansion produces condensation on the particles, which act as condensation nuclei. The cloud of condensed droplets attenuates the light beam resulting in an appropriate output signal relative to particle concentration.

Expansion and flushing out of the sample is rapid so overall measurement cycles are typically about one second long. This noncontinuous flow makes it necessary to incorporate an antipulsation device between diffusion batteries and the counter when it is used for this purpose, because flow must be continuous and uniform through the diffusion batteries. Dilution or bypassing of gases may also be necessary to mate counters to diffusion battery requirements.

6.5 OPTICAL COUNTERS

Optical or photoelectric counters function on the principle of light scattering. Reemission of light energy (fluorescence) by contrast can be used for chemical analyses. The counters have an upper limit of from about 300-10,000 particles/cm^3. Some can effectively measure particles 0.05-40 μm in size. Particle index of refraction is significant for particles $<$ 2 μm, and amount of scattered light is proportional to particle size for larger particles. Different types of particles in the 0.4-1 μm size range can cause measurements to vary by as much as two-fold. A discussion of optics and particle light scattering is not included as this is beyond the scope of this work.

Combination of various devices as shown in Figure 6.8 by Smith[14] enabled them to accurately measure particle sizes ≥ 0.005 μm in size. The cyclone precollector is used at the front of the sampling train to remove larger particles (d_{50} $\cong$ 2 μm) to reduce plugging. An antipulsing chamber is used between the diffusion battery and the condensation nuclei (CN) counters.

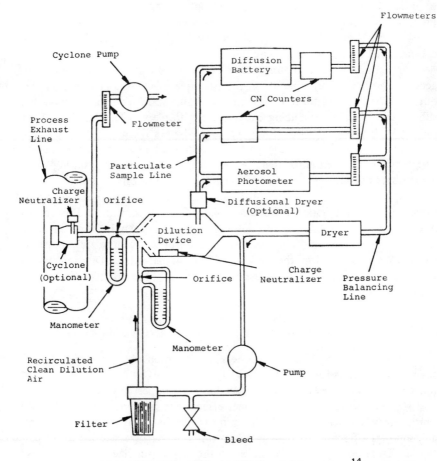

Figure 6.8 SRI optical and diffusional sizing system.[14]

6.6 ELECTRICAL ANALYZERS

Numerous electronic instruments are available to aid in the increasingly important particle size distribution and mass concentration measurements. A few of these systems are mentioned here to show what can and is being done; technical details related to the electronics and function of the systems are not given. Several of the devices are being used to electrostatically charge particles either to divide them for size distribution evaluation or maintain them completely mixed so *no* separation occurs. The first technique consists of charging the particles positively by corona discharge. The analyzer section of the unit, a negatively charged metal rod, attracts some of the particles. A low charge on the rod causes small particles, which have a high electrical mobility, to move to the rod during their passage through the analyzer. Those not collected on the rod pass through a current-collecting filter where the drain-off of electrons is indicated on an electrometer. A stepped increase in voltage on the rod causes correspondingly larger particles to be attracted, resulting in a decreased current on the electrometer. This is related to number of particles of the size determined by the rod voltage increment. A total of 11 voltage steps separate particles ranging in size from 0.0032-1.0 μm.

Particles larger than about 1 μm cannot be separated because their electrical mobility does not vary significantly. Particles smaller than 0.01 μm have high electrical mobility differences, but each particle accepts no more than a single charge and only a few of these particles can be charged. Because of this, a large change in numbers of particles causes only a small current change. Liu and Pui[28] report that at a space-time charge of 10^7 ion sec/cm^3, 0.1 μm particles acquire an average of 3.3 unit charges per particle. This means most particles have three or four charges and a few have two or five charges. The charge variations per particle cause electrical mobility variations and hence, collection for specific voltages occurs at different locations for similar-sized particles. This degrades the sensitivity of the device.

Another type of electrostatic sampler charges the aerosol then prevents size classification by inducing a pulsed electrical precipitating field. Samples passed through this system can be deposited uniformly on a flat collecting surface. This is of particular value in obtaining representative quantitative samples on microscope slides and electron microscope grids.

Mass concentration, but not size distribution, of particulate matter can be directly and quickly obtained by commercially available electronic instrumentation. An example is the respirable aerosol mass monitor that measures the mass of the 0.01-3.5 μm particles present in just seconds. The device functions by using an electrostatic precipitator to charge particles which are then deposited on a piezoelectric microbalance sensor.[29] The particles > 3.5 μm are removed by an impactor before charging.

Mass of deposited particles on the quartz crystal sensor causes the natural oscillation frequency to decrease by an amount proportional to the mass. A timed frequency deviation is indicated and concentration is obtained from gas flow rate.

6.7 AEROSOL GENERATION

Collection theories, size measurement techniques and devices are checked using actual particle samples. The ideal sample is highly monodisperse with a known particle size, is spherical, has a unit density, contains no electrostatic charge, does not absorb moisture (which could change size and mass) is rigid, does not dissolve or react with any liquids or system components and can be produced in a highly stable, continuous fashion.

Aerosol generators developed for the purpose of providing test sample particles range from relatively simple devices which atomize and dilute suspensions of purchased spherical particles, to complex electronic systems that routinely produce streams of aerosols.

6.7.1 Collision Atomizer

Liquid suspensions of fine particles of various substances can be purchased and atomized by a Collision Atomizer to form test aerosol particles. A wide range of particle compositions are available including polystyrene, polyvinyltoluene, glass, carbon black, stainless steel, silver, aluminum, nickel, tungsten and other metals, metallic compounds and clays plus natural pollens and spores. Sizes of these materials are ≥ 0.09 μm with standard deviation of less than 0.01 μm. They are available as 10-30% solids suspended in a liquid. The polystyrene particles have a density of 1.05 g/cm^3 and an index of refraction of 1.592 at 20°C.

Polystyrene latex particles 0.5-2.0 μm in 10% solutions were used by Calvert et al.[17] in testing cascade impactors. The Collision Atomizer is shown in Figure 6.9, and the entire calibration system is given as Figure 6.10. With this type of atomizer, it is necessary to periodically check the output stream by catching the aerosols on a slide and viewing them with a microscope to see that the number of double spheres does not exceed about 2%. If they do, further dilution of the liquid suspension is required. Once operating conditions have been established, Calvert reports that outlet concentrations and drop size are consistent for up to three hours if liquid level and concentration are maintained. Operating characteristics for this type of atomizer as given by May[30] are listed in Table 6.6.

About 45 l/min of dry filtered air is mixed with the atomizer exit stream to condition the drops and reduce agglomeration. This also helps develop

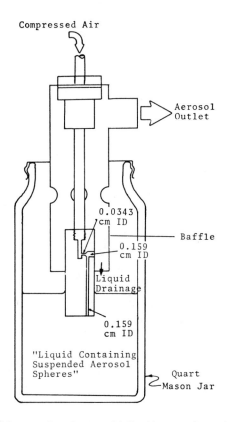

Figure 6.9 Collision atomizer for use with liquid suspensions of solid aerosols.[17]

desired exit particle concentration. Calvert then dried the drops using a 1.2-m long by 3.8-cm diameter horizontal glass tube with about a 1.5-cm layer of silica gel. Radioactive alpha emitters at the end of this tube neutralized any particle charges.

A disadvantage of these systems is that they are limited to the sizes and types of particles available. Also, evaporated liquid droplets containing dissolved material produce residue particles, which can result in a polydisperse test aerosol.

6.7.2 Aerosol Generators

These units create the test particles and release them into a test gas stream. Pneumatic atomizers can do this by atomizing a solution of a soluble substance and evaporating the solvent to form solid particles, or by simply atomizing a liquid to form liquid particles. The size of solid particles can

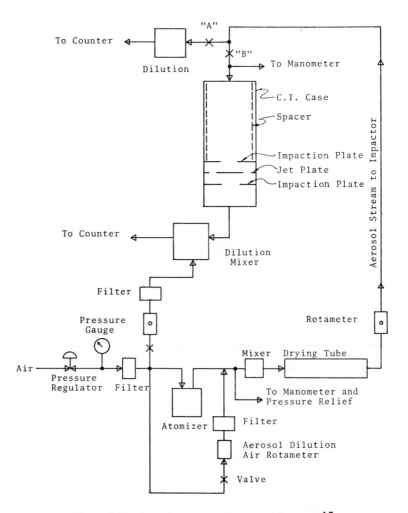

Figure 6.10 Cascade impactor size calibration train.[17]

be varied both by changing the solution concentration and changing the atomized droplet size. This procedure can be a disadvantage because during operation the solvent evaporates, causing concentration and therefore particle size changes.

Liu and Lee[31] have attempted to reduce the problem of solvent evaporation and have developed a constant-liquid feed atomizer. Their atomizer is similar to the Collision Atomizer and is shown in Figure 6.11. Air at 180 cm Hg gauge passes through the 0.0343-cm orifice and atomizes the liquid

Table 6.6 Operating Characteristics of a Three-Jet Collision Atomizer

Air pressure, cm Hg gauge	77.3	103.5	129.1	155.3	208.1	258.9
Air required, ℓ/min	6.1	7.1	8.2	9.4	11.4	13.6
Water vapor loss, approx. mℓ/hr	4.6	5.4	6.2	7.1	8.6	10.2
Water droplet loss, approx. mℓ/hr	3.2	3.3	3.3	3.3	3.4	3.8
Water concentration in outlet, g/m^3	21.3	20.4	19.3	18.4	17.5	17.2

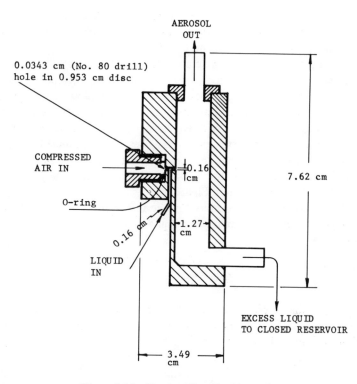

Figure 6.11 Constant-flow liquid atomizer.

introduced through the 0.16-cm tube. The coarse liquid droplets formed impact on the far vertical wall and drain down while the fine spray aerosol goes out the top of the unit. Gas and liquid flow rates and liquid concentration can be varied independently to produce the aerosol, which can then be diluted, dried and neutralized. Dioctylphthalate (DOP) has been used in this system to produce drops with median diameters from 0.03-1.3 μm and standard deviations of 1.4 and 1.2, respectively.

The spinning disc generator with a constant-liquid solution feed onto a spinning disc eliminates the problem of solution concentration change. The sizes produced by such devices are larger; e.g., 1-10 μm particles from a DOP-ethyl alcohol mixture with standard geometric deviations from 1.05-1.15. Very fine droplets falling close to the edge of the disc enter a separate chamber and are removed from the test aerosol. The primary droplets that fall beyond this have a size predicted by

$$d = C_1 \left[\frac{\Upsilon}{\rho_p \, \omega^2 \, D_D} \right]^{\frac{1}{2}} \qquad (6.20)$$

where Υ is liquid surface tension, ω is disc angular speed, D_D is disc diameter and C_1 is a constant. Theory gives C_1 as 3.5, but experimentally it ranges from 2-7 depending on disc speed and liquid used.

A vibrating-orifice aerosol generator developed by Berglund and Liu[31] is used to generate particles from 0.6-40 μm with average geometric standard deviation of 1.06. This device causes a cylindrical liquid jet to be broken up by a constant-frequency mechanical disturbance of sufficient amplitude to produce equal-sized droplets. Different size streams (from different size orifices) increase the droplet size spectrum. These droplets must be dispersed and diluted as soon as formed to prevent coagulation. A sketch of this generator showing the orifice disc (3-22 μm diameter) and the piezoelectric ceramic crystal vibrator is given in Figure 6.12.

The liquid to be atomized must first be in the form of a jet. Lindblad and Schneider[32] derived the minimum liquid velocity to produce a jet from a capillary tube:

$$v_{min} = \left(\frac{8 \, \Upsilon}{\rho_\varrho D_j} \right)^{\frac{1}{2}} \qquad (6.21)$$

where D_j is jet diameter. The optimum wavelength (distance between disturbances), λ_{opt}, for the break-up of an inviscid, incompressible, infinitely long cylindrical liquid jet as developed by Rayleigh[33] is

$$\lambda_{opt} = 4.508 \, D_j \qquad (6.22)$$

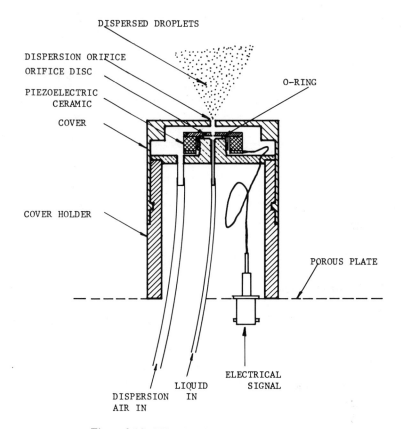

DISPERSED DROPLETS

DISPERSION ORIFICE
ORIFICE DISC

PIEZOELECTRIC
CERAMIC

COVER

O-RING

COVER HOLDER

POROUS PLATE

ELECTRICAL
SIGNAL

LIQUID
IN

DISPERSION
AIR IN

Figure 6.12 Vibrating-disc aerosol generator.

Schneider and Hendricks[34] showed that uniform droplets could be produced by varying λ from 3.5 to 7 D_j.

Size of aerosol produced by the vibrating-orifice generator can be estimated by

$$d = \frac{10^4 (6\ C_s\ Q_\varrho)^{1/3}}{\pi\ f} \tag{6.23}$$

where C_s is volumetric fraction concentration of the nonvolatile solute, Q_ϱ is liquid flow rate in cc/min and f is vibration frequency in KHz. If solid particles are to be produced by evaporation of droplets, their size is much smaller depending on the value of C_s. Presence of nonvolatile impurities in the liquid results in increased error from particle size predicted by Equation (6.23) as smaller particles are produced. Typical operating and particle production values, using optimum solids concentration of DOP in the liquid, are given in Table 6.7 for this type of system. Normal dilution, drying and neutralization procedures would be used in the final aerosol preparation.

Table 6.7   Operating Conditions and Production in a Vibrating-Orifice Particle Generator.[31]

Orifice Diameter (μm)	Liquid Flow (cc/min)	Frequency (KHz)	Droplet Range (μm)	Ultimate Minimum Particle Size (μm)	Typical Particle Concentration (no./cm^3)
22.1	0.214	70	44.0-49.8	1.19	40
8.25	0.0623	450	14.7-18.9	0.399	254
2.96	0.0220	725	9.21-10.8	0.250	410

6.7.3 Ultrafine Particles

Ultrafine particles can be formed by condensation of fumes and by exposure of gases to energy such as photolytic, radiolytic, pyrolytic and x-irradiation processes. In the case of these very fine particles, particle *diameter* is reported to affect the properties as much and even more than the *chemical composition* of the species. Surface area per unit mass, A_s, and diameter of ultrafine particles are related by

$$d = \frac{6}{\rho_p\, A_s} \tag{6.24}$$

Nuclei of condensation can be made to appear in atmospheric air that has been thoroughly filtered to remove all particles. These ultrafine particles are formed, for example, from gaseous impurities as a direct result of the intensity of solar radiation. Particles as large as 0.01 μm formed by this process were observed, but this was probably the result of coagulation.

Metal oxide particles in the 0.01-0.2 μm range are formed by pyrolysis of metal chloride vapors in an oxygen-hydrogen flame. Formenti *et al.*[35] report that particle geometry and crystalline structure depend on flame temperature flow rate of carrier gas and velocity of the metallic chloride vapors. The properties of ultrafine particles differ from that of larger particles, as noted in Chapter 1.

Spherical, submicron particles are produced by x-irradiation of organic vapors. These particles have a color characteristic of the material from which they are produced and are usually nonvolatile and very stable. The size and concentration of particles formed by King *et al.*[36] depended on the organic material used, vapor concentration, humidity, X-ray intensity, X-ray time and aging time. Particle formation continues for a while after the X-rays are discontinued, if organic material remains present.

REFERENCES

1. Brooks, E. F. and R. G. Williams. "Flow and Gas Sampling Manual," EPA-600/2-76-703 (July 1976).
2. "Performance and Measurements at Dust Collectors," Verein Deutscher Ingenieure, VDI-2066 Standards (May 1966).
3. Oglesby, S., Jr. and G. B. Nichols. "A Manual of Electrostatic Precipitator Technology, Part I," Southern Research Institute (August 1970).
4. Hoffman, W. *VDI-Ber* 7(15) (1955).
5. Meyer, S. L. *Data Analysis for Scientists and Engineers* (New York: John Wiley & Sons, Inc., 1975).
6. May, K. R. "The Cascade Impactor: An Instrument for Sampling Course Aerosols," *J. Sci. Inst.* 22 (October 1945).
7. Brink, J. A., Jr. "Cascade Impactor for Adiabatic Measurements," *Ind. Eng. Chem.* 50(4) (1958).
8. Anderson, A. A. "New Sampler for the Collection, Sizing, and Enumeration of Viable Airborne Particles," *J. Bacteriol.* 76 (1958).
9. Rantz, W. E. and J. B. Wong. "Impaction of Dust and Smoke Particles," *Ind. Eng. Chem.* 44(6) (1952).
10. Cohen, J. J. and D. N. Montan. "Theoretical Considerations, Design and Evaluation of a Cascade Impactor," *Am. Ind. Hyg. Assoc. J.* March-April 1967).
11. Lundgren, D. A. "An Aerosol Sampler for Determination of Particle Concentration as a Function of Size and Time," *J. Air Poll. Control Assoc.* 17(4) (1967).
12. Pilat, M. J., D. S. Ensor and J. C. Busch. "Cascade Impactor for Sizing Particles in Emission Sources," *Am. Ind. Hyg. Assoc. J.* 32(8) (1971).
13. Gussman, R. A., A. M. Sacca and N. M. McMahon. "Design and Calibration of a High Volume Cascade Impactor," *J. Air Poll. Control Assoc.* 23(9) (1973).
14. Smith, W. B., K. M. Cushing and J. D. McCain. "Sizing Techniques for Control Device Evaluation," SORI-EAS-74-138, Southern Research Institute (July 12, 1974).
15. Mercer, T. T. and R. G. Stafford. *Am. Occup. Hyg.* 12:41 (1969).
16. Calvert, S. *et al.* "Scrubber Handbook," APT, Inc. (1972).
17. Calvert, S., C. Lake and R. Parker. "Cascade Impactor Calibration Guidelines," EPA-600/12-76-118 (April 1976).
18. Pollution Control Systems Corp. "Operation Manual Mark III University of Washington Source Test Cascade Impactor, Model D" (March 1974).
19. Raabe, O. G. "Aerosol Aerodynamic Size Conventions for Inertial Sampler Calibration," *J. Air Poll. Control Assoc.* 26(9):856 (1976).
20. Smith, W. B., K. M. Cushing and G. E. Lacey. "Anderson Filter Substrate Weight Loss," EPA-650/2-75-022 (February 1975).

21. Hesketh, H. E.   "Aerosol Capture Efficiency in Scrubbers,"   paper 75-50.6 presented at the 68th Annual Air Pollution Control Association meeting, Boston, June 1975.
22. Picknett, R. G.   "A New Method of Determining Aerosol Size Distributions from Multistage Sampler Data," *Aerosol Sci.* (1972).
23. Townsend, J. S.   *Trans. Roy. Soc.* (London) 193A:129 (1900).
24. Gormley, P. G. and M. Kennedy.   "Diffusion from a Stream Flowing Through a Cylindrical Tube," *Proc. Roy. Irish Acad.* 52A:163 (1949).
25. Thomas, J. W.   "The Diffusion Battery Method for Aerosol Particle Size Determination," *J. Coll. Sci.* 10:246 (1955).
26. Sinclair, D. and G. S. Hoopes.   "A Novel Form of Diffusion Battery," *Am. Ind. Hyg. Assoc. J.* (January 1975), p. 39.
27. Sinclair, D., R. J. Countess, B. Y. H. Liu and D. Y. H. Pui.   "Experimental Verification of Diffusion Battery Theory," *J. Air Poll. Control Assoc.* 26(7):661 (1976).
28. Liu, B. Y. H. and D. Y. H. Pui.   "On the Performance of the Electrical Aerosol Analyzer," Pub. No. PTL-237, Particle Technology Laboratory, Department of Mechanical Engineering, University of Minnesota, Minneapolis (1975).
29. Sem, G. J.   "Piezobalance Respirable Aerosol Sensor: Application and Performance," American Industrial Hygiene Association Annual Meeting, May 1976.
30. May, K. R.   "The Collision Nebulizer: Description, Performance and Application," *Aerosol Sci.* 4:235 (1973).
31. Berglund, R. N. and B. Y. H. Liu.   "Generation of Monodisperse Aerosol Standards," *Environ. Sci. Technol.* 7(2):147 (1973).
32. Lindblad, N. R. and J. M. Schneider.   *J. Sci. Inst.* 42:635 (1965).
33. Rayleigh, L.   *Proc. Roy. Soc.* 29:71 (1879).
34. Schneider, J. M. and C. D. Hendricks.   *Rev. Sci. Inst.* 35(10):1349 (1964).
35. Formenti, M. *et al.*   "Preparation in a Hydrogen-Oxygen Flame of Ultrafine Metal Oxide Particles," In *Aerosols and Atmospheric Chemistry* G. M. Hidy, Ed. (New York: Academic Press, 1971).
36. King, E. Y. H. *et al.*   "Aerosols Produced by X-Rays," In: *Aerosols and Atmospheric Chemistry,* G. M. Hidy, Ed. (New York: Academic Press, 1971).

CHAPTER 7

EXPERIMENTS AND PROBLEMS

7.1 EXPERIMENTS

Commercial equipment can be purchased to measure particle behavior and to obtain data relevant to particle emissions and collection. However, it is useful to have a working knowledge of several basic devices. This should improve the quality of the data by helping to ensure the correct use and calibration of the devices and the system. In addition, there are times when lack of funds or an emergency situation may make it necessary to improvise equipment to enable a test to be performed.

This chapter focuses on construction of several basic items used for obtaining data related to particulates. It gives few details, thus it is suggested that the readers use their own imaginations to expand and improve on these ideas. All normal testing procedures must be incorporated as necessary to obtain complete and useful data. Laboratory balances, dessicators and other equipment are needed in addition to items specified.

7.1.1 Impingers, Limiting Flow Orifice and Manometers

An impinger train and limiting flow orifice can be constructed and calibrated using many common laboratory items. An impinger train can be used to obtain samples of gaseous pollutants (by absorption) or for collection of particulate matter (by impaction). Construct a train of three midget impingers and assemble as shown in Figure 7.1.

Calibrate each impinger for from 5-30 mℓ volume as shown in Figure 7.2(a). (Read bottom of the water meniscus.) The impinger nozzles should

---

SAFETY NOTE: Break glass properly to prevent cuts; do not force glass (*e.g.*, into stopper holes, lubricate first); hot glass is *hot*; get instructions if necessary. Wear safety glasses!

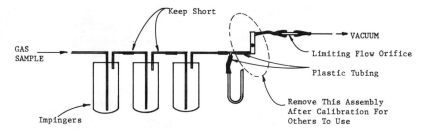

Figure 7.1   Impinger train and limiting flow orifice.

be close to ~ 1 mm inside diameter (ID). Prepare by flame pulling molten glass tubing, break off and measure the opening with a 1-mm diameter drill. If tubing end is pulled too small, sand carefully on emery paper or rework in flame. Orifice tip size can be reduced by heating in the flame gently while turning. It may be necessary to simultaneously blow through the tube when preparing small orifices.

A limiting flow orifice can be constructed by pulling 7-mm OD glass tubing to obtain an opening of approximately 0.37 mm and placing it in the line as shown in Figure 7.2(b). Check size with a drill. Hypodermic needles can also be used for this. A proper orifice should limit air flow to 1-2 ℓpm when a vacuum is applied. Check flow with a flow meter and record value. If a ball and tube meter is used, read flow at the *center* of the ball.

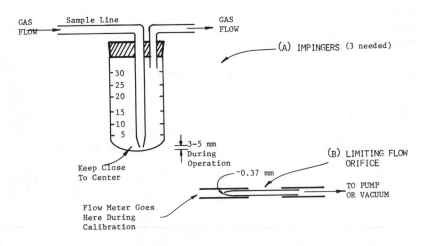

Figure 7.2   Details of impingers and limiting flow orifice.

Construct a "U" tube manometer for about 40 cm H_2O ΔP, fill with colored water and make a calibration scale. An inclined manometer ($\sim 15°$) is often advantageous for increasing the accuracy of pressure drop readings. Construction details for this are given at the end of this experiment. Colored water can be made using a few drops of vegetable dye per 500 mℓ distilled water.

Assemble the train as shown in Figure 7.1 and calibrate your flow orifice. During calibration, measure flow rate, pressure and temperature. The flow meter is removed after calibration and the system is now ready for field use. The system is limited to these specific impingers, orifice and liquid levels if liquid is used. Any change in these requires recalibration.

Discuss:

1.   The principle of operation of a limiting flow orifice;
2.   Why an impinger can work for (a) gaseous pollution removal and (b) particulate capture; and
3.   Why ΔP values and liquid quantities in each impinger should be recorded during calibration.

Supplies needed are:

1.	7 mm glass tubing	10.	Burners
2.	1/4-in. ID vinyl tubing	11.	Vegetable (or other) dye
3.	3-25 x 150 mm test tubes	12.	Stopper hole drills
4.	Distilled H_2O	13.	Marker
5.	Emery paper	14.	Masking tape
6.	File	15.	Rule
7.	Graduated cylinder	16.	Small drills (0.37 mm = #79)
8.	Calibrated flow meter		(1 mm = #61)
9.	Rubber stoppers	17.	Vacuum source

Procedure for Constructing and Using Homemade Inclined Manometer: Obtain a piece of scrap plywood or cardboard with straight sides about 30 x 50 cm. Cut or mark sides so they are square ($90°$) with top and bottom. Bend glass tubing so it has some desired angle ($\sim 15°$ from horizontal). Using flexible tubing, attach a plastic bottle or other container as a "large" water reservoir as shown in Figure 7.3. Mount the vertical side perpendicular to bottom and top. Then fill to desired level with colored water.

Assume glass tubing is uniform in ID over the range manometer is to be used (a good assumption except where bend occurs). Secure a piece of graph paper to the board so the lines are parallel with bottom edge as shown in the figure. Measure distances along ordinate and abscissa of the manometer as shown by the example in Figure 7.3. The example values show the ΔP multiplication advantage of this gauge. The example is 45/20 = 2.25.

Construct a scale on a clear plastic card. In the example, a distance of 2.25 cm should be labeled 1 cm, 4.5 cm as 2 cm, etc. as shown in Figure 7.4.

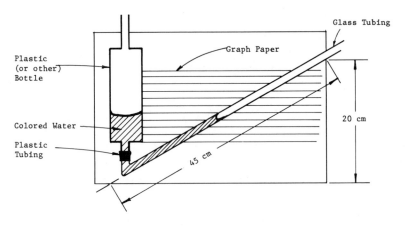

Figure 7.3   Inclined manometer.

Subdivide the card to read 0.2-cm units if possible. Slide this card under the inclined glass side and leave loose.

To read manometer, level top (or bottom) of board. Note bottom of meniscus on vertical leg, and locate that point on inclined leg using the parallel lines on graph paper. Slide plastic card so 0 is at this point. Read ΔP in cm of water directly from plastic calibration scale at bottom of upper meniscus.

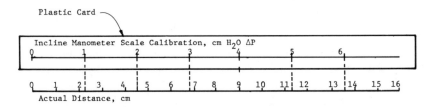

Figure 7.4   Sample incline manometer calibration scale.

7.1.2 Determination of Particulate Concentration

Normal test procedures are not discussed but must be incorporated as necessary to obtain useful data.

7.1.2.1 Impinger Train

The impinger train can be used for this with either about 10 cc of distilled water in the first impinger or with all dry. It would be necessary to evaporate the water and/or weigh all before and after tests to obtain total particulates captured. In addition, a final filter must be used *after* the orifice to collect any very fine particles.

7.1.2.2 High Volume Sampler

An alternate procedure for atmospheric particulates is to make a high-volume arrangement from a vacuum cleaner with an attachment hose. Depending on the sweeper air flow rate, either the hose end (about 4 cm in diameter) or an attachment on the hose can be used for the filter holder. It has been shown that several layers of inexpensive-grade toilet tissue can give the desired filter resistance. The paper must be dried and weighed then secured to the end of the hose; for example, using a rubber band. The sweeper air flow rate, with a filter of the type to be used for the test in place, can be calibrated by timing the filling of a large plastic bag or other suitable object and calculating the volume. The bag should be only partially filled. Can you explain?

7.1.2.3 Visual Emissions

A crude estimate can be made of emissions using a chart as suggested by Professor Maximilian Ringelmann of Paris in about 1888.[1] The chart can be made by placing black lines on white cards to obtain five equal steps between black and white:

    Chart 0 – all white
    Chart 1 – 1-mm thick black lines, 10 mm apart leaving 9-mm white squares
    Chart 2 – 2.3-mm thick lines, 7.7-mm white square
    Chart 3 – 3.7-mm thick lines, 6.3-mm white square
    Chart 4 – 5.5-mm thick lines, 4.5-mm white square
    Chart 5 – all black

The charts should be about ≥ 22 cm in size.

The charts are held level with the eye and the plume to be observed, at a distance such that the lines on the chart merge into shades of grey. The chart that appears the same shade as the smoke gives the smoke shade number. A clear stack would be #0 and a pure black smoke is #5. Estimates of half shade are also used when needed. Background lighting, smoke color, condensed moisture and other factors influence the plume appearance.

Average smoke density can be obtained after numerous readings are obtained (*e.g.,* one hour). This is obtained by adding the number of observations of each shade and multiplying that value times the respective shade number. Sum these totals to obtain a grand total of the readings and obtain

$$\% \text{ smoke density} = \frac{(\text{grand total}) \, (20)}{\text{total no. of individual observations}}$$

7.1.2.4 In-Line Filter

A simple filter can be constructed for obtaining particulate mass measurements using a cellulose thimble, glass jar, glass tubing and rubber stopper, as shown in Figure 7.5. The large stopper has two holes so one tube can fit

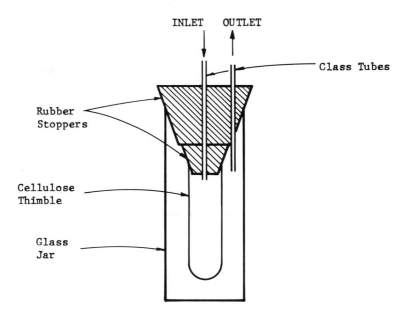

Figure 7.5   Thimble holder in-line filter.

outside the thimble and one tube can fit through a smaller stopper and into the thimble. The thimble is pushed securely onto the smaller stopper so the particles are captured inside the thimble. A limiting flow orifice should be appropriately sized and calibrated to provide the desired sampling velocity for isokinetic sampling with the particular probe size used and system gas velocity. The test probe can be made by bending a piece of glass tubing at a 90° angle to obtain an "L" shape. The end of the "L" should be about 10 cm long, and rough edges should be removed by fire polishing.

7.1.3 Gas Measurements

A pitot tube to provide gas velocity profiles and duct dimensions to give cross-sectional area can be used to obtain volumetric gas flow data.

Temperature, pressure and water content data are needed to complete the information for gases that approximate air. Composition, viscosity and density information are needed in addition for other gases.

Standard pitot tube operation and use is given in Section 6.1.2. These devices, if taken care of, give accurate readings from which velocity can be calculated. However, the standard pitots require more elaborate construction facilities. By contrast, a Type S or Stauscheibe pitot can be made easily, but it is necessary to calibrate it against a standard pitot or other calibra-tioned system.

A Type S pitot is shown in Figure 7.6. Construct the tube using metal or glass tubing and wire, tape, etc. The pitot ends must face 180° away from each other and the openings must be in parallel planes. Center the upper opening directly above the lower. A manometer, as constructed in Section 7.1.1, is needed to indicate pressure drop. The pitot end facing upstream indicates static plus kinetic gas pressure and the end facing downstream reads static pressure.

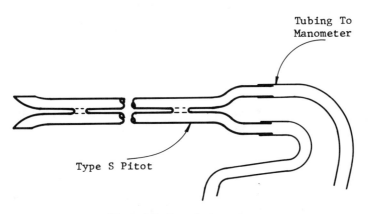

Figure 7.6 Type S pitot tube.

Supplies needed are:

1. About 6 ft of 1/4-in. copper, glass (6 or 7 mm) or other tubing
2. Wire and/or solder
3. Burner
4. File, tubing cutter, and emery paper
5. Rubber bulb
6. 100 mℓ graduated cylinder
7. Pinch clamps
8. Tubing
9. Standard pitot
10. Blower and air tube
11. 1/4-in. ID vinyl tubing

7.1.3.1 Velocity

All pitot tubes are directional; however, the S type should be calibrated and used with *one specific* end forward. Both ends could be calibrated and the appropriate calibration used for whichever side is forward, but this could lead to confusion and error. The standard pitot is symmetrical and directional orientation depends on only one angle–the tilt angle. Orientation of the S tube depends on two angles–the rotation angle, θ_1, as shown in Figure 7.7(a) and the tilt angle, θ_2, shown in Figure 7.7(b).

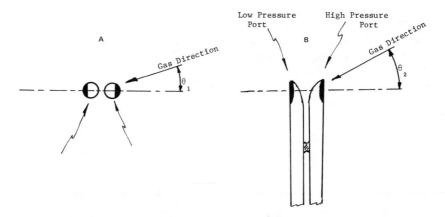

Figure 7.7   S-type pitot tube. (a) end view to show rotation orientation; (b) side view to show tilt orientation

Using the pitot constructed, show it is a directional device by taking readings of ΔP at the center of a gas duct at $0°$ tilt and rotation, and at $5°$ intervals from 0-$30°$ rotation and the same for tilt. Plot these data as ΔP versus angle to show relative error with respect to the value at $0°$.

The pitot must be calibrated against a standard pitot or against another pitot that has been standardized. The correction factor of the pitot being tested, C_{test}, is found by

$$C_{test} = C_{std} \sqrt{\frac{\Delta P_{std}}{\Delta P_{test}}} \qquad (7.1)$$

where the C_{std} equals 1.0 for a standard pitot or the appropriate value for the standardized pitot. Correction factors for S tubes are usually about 0.85. This factor can vary with gas velocity.[2]

Velocity can now be determined for air using

$$v = 1495\, C_{test} \sqrt{\frac{T}{273.16}} \sqrt{\Delta P} \qquad (7.2)$$

where v is in cm/sec, T is degrees Kelvin and ΔP is mm Hg. The velocity of other gases can be found by Equation (6.2), remembering the pitot correction must be applied when necessary.

Measure the velocity profile in two planes of a duct using the pitot and calculate volumetric flow rate at standard conditions. Duct pressure, temperature and average $\sqrt{\Delta P}$ (or average of individual velocities) are needed to do this. Take 12 readings at centers of equal areas as shown by Figure 7.8(a) or (b) and Table 7.1.[3] Twelve sample points are adequate if the location is eight duct diameters or more downstream from last disturbance and two diameters or more upstream from any disturbance.

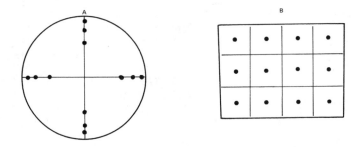

Figure 7.8 Equal area sample locations. (a) cross section of circular stack divided into 12 equal areas, showing location of traverse points at centroid of each area; (b) cross section of rectangular stack divided into 12 equal areas, with traverse points at centroid of each area.

7.1.3.2 Pressure

Determine duct static pressure by (a) using your pitot and (b) using a sampling port opening. The S pitot turned 90° reads static pressure on *either* side. A connection in the duct flush with the inside wall gives static pressure.

7.2 PROBLEMS

Several typical problems are given below to suggest the type of calculations that can be made. Some problems require extensive time to solve completely.

1. Consider a liquid particle in free fall in still air at SC; d = 3.6 x 10⁻⁵ cm; $\rho_p = 1.0$ g/cm³
   a. Establish the size regime
   b. Obtain C
   c. Determine v_s by two different methods
   d. Find F_D

Table 7.1 Location of Traverse Points in Circular Stacks[3]
(Percent of Stack Diameter from Inside Wall to Traverse Point)

Traverse Point Number on a Diameter	Number of Traverse Points on a Diameter											
	2	4	6	8	10	12	14	16	18	20	22	24
1	14.6	6.7	4.4	3.3	2.5	2.1	1.8	1.6	1.4	1.3	1.1	1.1
2	85.4	25.0	14.7	10.5	8.2	6.7	5.7	4.9	4.4	3.9	3.5	3.2
3		75.0	29.5	19.4	14.6	11.8	9.9	8.5	7.5	6.7	6.0	5.5
4		93.3	70.5	32.3	22.6	17.7	14.6	12.5	10.9	9.7	8.7	7.9
5			85.3	67.7	34.2	25.0	20.1	16.9	14.6	12.9	11.6	10.5
6			95.6	80.6	65.8	35.5	26.9	22.0	18.8	16.5	14.6	13.2
7				89.5	77.4	64.5	36.6	28.3	23.6	20.4	18.0	16.1
8				96.7	85.4	75.0	63.4	37.5	29.6	25.0	21.8	19.4
9					91.8	82.3	73.1	62.5	38.2	30.6	26.1	23.0
10					97.5	88.2	79.9	71.7	61.8	38.8	31.5	27.2
11						93.3	85.4	78.0	70.4	61.2	39.3	32.3
12						97.9	90.1	83.1	76.4	69.4	60.7	39.8
13							94.3	87.5	81.2	75.0	68.5	60.2
14							98.2	91.5	85.4	79.6	73.9	67.7
15								95.1	89.1	83.5	78.2	72.8
16								98.4	92.5	87.1	82.0	77.0
17									95.6	90.3	85.4	80.6
18									98.6	93.3	88.4	83.9
19										96.1	91.3	86.8
20										98.7	94.0	89.5
21											96.5	92.1
22											98.9	94.5
23												96.8
24												98.9

    e. Calculate the value of B
    f. Determine Re_p

2. **Repeat** # 1 using a particle with d = 0.012 μm and a particle with d = 120 μm.

3. If the particle in # 1 were falling at 10^{-4} cm/sec, what is the diameter?

4. Determine the terminal settling velocity of a cubical particle 2 μm on a side with a ρ_p = 1 g/cm^3 and in still air at 5°C. Give the equivalent and sedimentation diameters for this.

5.   Repeat # 4 for a chain of glass beads consisting of a flat aggregate of three beads. Each bead has a 3 μm diameter.

6.   It is observed that a glass particle falls in the air at a rate of 0.057 cm/sec under influence of gravity and an electric potential. Reversed potential causes the particle to fall at 0.005 cm/sec. The particle is nonspherical. The electrical plates are spaced 2 cm apart, and the voltage needed to suspend the particle is 14.9 volts. Find:

    a. Terminal settling velocity
    b. Particle charge in electrons
    c. Particle mobility (electrical and mechanical)
    d. What voltage was used to obtain the electrical velocity plus sedimentation velocity value?

7.   A particle of liquid with a ρ_p = 1 g/cm^3 is in stationary air at SC. For a 2-μm diameter particle, determine at 10 μsec after start of free fall:

    a. Velocity
    b. Acceleration
    c. Distance traveled

8.   For a 2-μm diameter particle as in # 7, determine at 10 μsec after gas expands horizontally into an infinitely large chamber from an initial velocity of 6000 cm/sec:

    a. Velocity
    b. Acceleration
    c. Distance traveled

9.   For a 100-μm diameter particle of p$_p$ = 1 and spherical, determine at 0.01 sec after start of free fall in air at SC (this can be solved by graphical integration):

    a. Velocity
    b. Acceleration
    c. Distance traveled

10. A 100-μm diameter particle of p$_p$ = 1 and spherical is moving with an air stream at SC. Determine at 0.01 sec after the gas expands horizontally into an infinitely large chamber from an initial velocity of 6000 cm/sec (solve by graphical integration):

    a. Velocity
    b. Acceleration
    c. Distance traveled

11. Tabulate for 0.012, 0.36 and 120 μm spherical particles with a density of 1.5 g/cm^3 in still air at SC the following:

    C, B, b, D_{PM}, τ, v_d, ℓ_d and $\overline{\Delta x_d}$.

12. Assume an inertial wet scrubber has the following representative scrubbing conditions: relative velocity difference is 10 m/sec, number of

drops is $10^9/m^3$, particle density is 1 g/cc, gas is air at SC, and size of collecting drops is 50 μm. Estimate particle scrubbing time, τ_{SC}, for:

    a. 1-μ particles
    b. 0.3-μ particles
    c. What is the relationship between these scrubber cleaning times for the different size particles?

13. Use the data from Problem 12 and determine the minimum size particle that can be effectively scrubbed by an inertial wet scrubber under these conditions. Assume drop density is also 1 g/cm^3.

14. Derive the relationship giving the ratio of CDS cleaning time to inertial wet scrubber cleaning time as a function of charged particle diameter; *i.e.,* = f(d)^3. Assume saturation charging of drops and particles due to impact charging in a charging field E_c.

15. For similar scrubbing conditions as in Problem 13, show that the collection performance of a conventional scrubber would be expected to be improved by charging the drops and the particles (use the relationship τ_c/τ_{SC}). Base this on 0.5-μm particles, 50-μm drops, 10 m/sec relative velocity difference, drop and particle densities both of 1 g/cm^3 and in air at SC. Assume a linear charging field is obtained by a 20 KV potential across a 2-cm gap.

16. A 1.25-cm ID tube is used to sample particulates from a power plant. Estimate what overall percent of the particulates are lost in a 3-m long section of horizontal tube assuming the gas is air at SC, particulate density is 1 g/cm^3 and sampling rate is 0.65 m^3/hr. For simplicity, base work on largest particle size of 60 μm, mean particle size of 6 μm and smallest particle size of 0.6 μm. Particles may be assumed to be spherical. What is the flow Reynolds number in the tube?

17. Midget impingers are constructed with a 1-mm ID orifice located about 3 mm from the bottom of a 26-mm ID receiver. Jet exit velocity is reported to be about 60 m/sec when sampling rate is 2.8 l/min max. at 30 cm H$_2$O vacuum. Greensburg-Smith impingers have a 2.3-mm ID orifice located about 5 mm from the strike or impingement plate. The ID of the receiver is 57 mm. At a sampling rate of 1.6 m^3/hr @ 7.6 cm Hg, the strike velocity is reported as 100 m/sec.

    a. Estimate the collection efficiency of each device on 1-μm particles in air at SC. The particles may be assumed to be rough spheres and have a density of 1 g/cm^3.
    b. Determine what size particles are captured with a 50% efficiency (this would give the "cut diameter").

18. A commercial cascade impactor (Model 4240) for atmospheric particle sizing designed by Dr. Dale Lundgren is arranged as shown by the top

view given in Figure 7.9. The width of the slit opening to stage #1 is 0.828 cm, to stage #2 is 0.249 cm, #3 is 0.089 cm and #4 is 0.031 cm. All slits are the same length (height), namely 4.763 cm. The impaction surface is a rotating drum (1 rpm to 1 rph) with 64.5 cm^2/stage.

    a. At a gas sampling rate of 6.4 m^3/hr, determine the cut diameter for particles on each stage assuming SC and spherical particles with a density of 1 g/cm^3.

    b. Calculate particle diameter collected with 90% efficiency for each stage. Neglect impactor pressure drop.

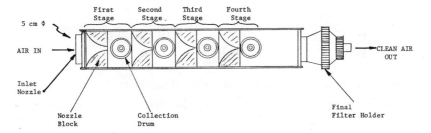

Figure 7.9 Lundgren cascade impactor.

19. An add-on "in-line" adapter of the type shown in Figure 7.10 is to be made for the Lundgren Cascade Impactor (see Problem 18) so it can be used as a source sampler. Specify the missing dimension that will produce the least detrimental influence on the operation of the impactor. List reasons for this decision and state assumptions. The emissions to be tested are expected to have a bimodal distribution. The mass mean diameter is 11.5 μm and the geometric standard deviation is 44.6 for the smaller 31% and 2.43 for the larger 69%. In addition, assume the particulates have a ρ_p of 1 g/cm^3 and are rough spheres, the gas is essentially air at 120°F, sampling rate is 6.4 m^3/hr, and inlet sample line is 2.22 cm ID.

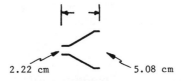

2.22 cm   5.08 cm

Figure 7.10 In-line adapter for converting atmospheric test impactor to source tester.

20. The Lundgren Cascade Impactor of Problem 18 is modified as in Problem 19 for stack sampling. Using the assumptions given in Problem 18 calculate the cut diameter for each stage if the gas is flue gas (M = 29.8) at 200°C. Remember to correct for gas viscosity assuming flue gas is essentially the same as air.

21. Find the approximate ratio of distance moved by diffusion to distance moved by gravity at SC for particles with a density of 2 g/cm^3 in air with (a) d = 0.10 μm, and (b) d = 10 μm over a 1 second period.

22. Estimate the diffusional velocity in cm/sec of a particle with a radius of 10^{-6} cm and a density of 2 g/cm^3 in air at (a) SC, and (b) 230°C.

23. A certain charge of fly ash (ρ_p = 1 g/cm^3) was measured as having the following percentage and specific charge:

   – positive 31% or 1.9 x 10^4 esu/g = 3.8 x 10^{13} electrons/g
   – negative 26% or 2.1 x 10^4 esu/g = 4.2 x 10^{13} electrons/g
   – neutral 43%

What can be said about particle size from these data?

24. Determine the ratio of adhesive force between a sphere and a plane to adhesive force between two equal-sized spheres. Consider the two spheres to be the same size as the single sphere and RH equal to 95%. Generalize how adhesion and RH influence ESP operation including particle removal from the collectors.

25. Establish adhesive force for 3-μm spherical particles and a 100-μm cylindrical fiber filter surface at 55% RH. List the effects of adhesion and RH on particle coagulation, fiber filtration, cake filtration and filter cleaning.

26. Spherical particles are being removed by a fabric filter with 100-μm cylindrical fibers at a filtering velocity of 1 cm/sec. Particle density is 2 g/cm^3 and the gas is air. Estimate collection efficiency of 0.01-, 0.1-, 1-, 10- and 100-μm particles collected by:

   a. Impaction
   b. Diffusion
   c. Direct interception
   d. Sedimentation

Plot these total efficiencies against particle diameter and note effective regimes for various collection mechanisms.

REFERENCES

1. "Ringelmann Smoke Chart," Information Circular IC 8333 U.S. Department of Interior, Bureau of Mines (May 1967).

2. Brooks, E. F. and R. L. Williams. "Flow and Gas Sampling Manual," EPA-600/12-76-203 (July 1976).
3. "Standards of Performance for New Stationary Sources," Method 1, 40 CFR 60; 36 FR 24876, *Federal Register* (December 23, 1971).

APPENDIX

TABLE OF NOMENCLATURE

LATIN

a = acceleration

A = surface area

b = electrical mobility of gas, ion or particle, equals v_e/E, cm^2/(V sec)

B = particle thermal (mechanical) mobility, equals v_d/F_d, sec/g

c = speed of light, equals 3×10^{10} cm/sec

c_A = molar concentration

c_P = gas specific heat, *e.g.*, cal/g $^\circ$K

c_T = thermal energy transfer constant, Eq. (4.49)

C = particle concentration by mass

C = various constants

C_f = final concentration

C_o = initial concentration

C_1 = constant, equals $4\, g\, \rho_p\, \rho_g\, d^3/3\mu^2 g$ in Eq. (3.12) and (3.18)

C = Cunningham correction factor, dimensionless

C_a = Cunningham correction factor using d_a, dimensionless

d = particle diameter, μm

$\overline{d}$ = mean diameter

d_a = aerodynamic diameter

d_{ar} = aerodynamic resistance diameter

d_A = molecular diameter of species A

d_c = cut diameter

d_e = equivalent diameter

d_o = projected diameter

d_s = sedimentation diameter

d_{50} = aerodynamic cut diameter, *i.e.*, particle diameter at which 50% collection efficiency occurs.

199

$d_{84.13}$ = diameter that occurs on a probability plot for a cumulative distribution of 84.13%

D = diameter, or equivalent, of containing device

D_c = diameter of collector, cm (also, effective cloud diameter)

D_i = rate of deposition of particle

D_j = diameter of jet

D_1 = diameter of collector, mm

d = differential operator

D_{AB}= diffusivity of species A in B

D_{PM}= diffusivity of particle in gaseous medium, cm^2/sec

D_t = eddy diffusivity of medium

e = natural logarithmic base, equals 2.718

e = one elementary charge, which is charge on 1 electron, equals 1.603×10^{-19} coulomb

E = external forces

E_o = overall collection efficiency, fraction unless otherwise noted

E = field strength, equals potential difference divided by distance between plates

E_p = precipitating field strength, max in air $\cong$ 10,000 V/cm

E_o = initial or charging field strength, max in air $\cong$ 10,000 V/cm

E_w = field strength at wall

f = frequency

F_a = adhesive force

F_B = buoyant force

F_d = diffusion force

F_e = electrostatic force

F_E = external forces acting on particle

F_D = total drag force

F_{D-1}= form drag force

F_{D-2}= friction drag force

F_{MA}= resistance forces of medium on accelerating particle

Fo = Fourier number, dimensionless

Fr = Froude number, dimensionless, see Eq. (1.20)

g = acceleration of gravity, $\sim$ 980 cm/sec^2 at sea level

g_c = Newton's Law of Motion conversion constant, 32.174 $lb_m ft/(lb_f sec^2)$

$\overline{G}^2$ = mean square particle displacement

h = distance between plates

h　　=　heat transfer coefficient, *e.g.,* cal/(sec-cm^2 °K)
H　　=　magnetic field intensity, oersted, equals 79.58 ampere turns per meter

I　　=　current, amperes

J_A　=　molar diffusion flux of A, g moles A/(cm^2-sec)

k　　=　Kozeny-Carman coefficient
k　　=　Boltzmann constant, equals 1.38 x 10^{-16} g cm^2/(sec^2 molecule °K)
k_g　=　thermal conductivity of gas
k_p　=　thermal conductivity of particulate matter
K　　=　filter bed permeability
K_1　=　filter cake-fabric resistance coefficient
K_I　=　impaction parameter, dimensionless, see St
Kn　=　Knudsen number, dimensionless, see Eq. (1.12)

ℓ　　=　distance from particle center to surface
L　　=　length of system
L　　=　liquid-to-gas ratio
L_{cr}　=　critical length to remove all particles

m　　=　meter
m_A　=　molecular mass of species A
m_p　=　mass of particle
M　　=　molecular weight, g/g mole
Ma　=　Mach Number, dimensionless, see Eq. (1.15)
MMD　=　mass mean diameter
M　　=　character of motion of cloud

n　　=　number of turns of conducting wire/cm
n_p　=　number of electrons on object
n_T　=　total number of particles deposited
N　　=　particle number concentration per cm^3
N_D　=　droplet number concentration per cm^3
N_g　=　initial concentration of gas per cm^3
N_o　=　initial concentration of ions per cm^3
N_p　=　initial concentration of particles per cm^3
N_{PD}=　particle diffusiophoresis flux rate
N_{PD}=　particle diffusion flux rate
N_{PT}=　particle thermophoresis flux rate
NA　=　numerical aperture

p_g = partial pressure of gas
p_v = partial pressure of vapor
P = total pressure
Pe = Peclet number
Pt = collector penetration, equals 1 - E, fraction
P° = saturation vapor pressure of pure liquid at a given temperature

q = charge on particle, equals n_pe
Q = volumetric flow rate
Q = charge on droplet

r = distance along radius of tube
R = universal gas constant, equals 82.05 atm cm^3/(g-mole $^\circ$K) or 83.14 x 10^6 g cm^2/(g-mole-sec^2 $^\circ$K)
R = radius of tube
Re_f = flow Reynolds number, dimensionless, see Eq. (1.9)
Re_p = particle Reynolds number, dimensionless, see Eq. (1.11)

s = distance
S = drag
S = specific surface area
Sc = Schmidt number, dimensionless, see Eq. (1.18)
St = Stokes number, dimensionless, see Eq. (1.19)
SC = Standard conditions, 20°C and 1 atmosphere

t = time
T = absolute temperature, $^\circ$K

$\bar{U}$ = bulk gas velocity

v = velocity of particle or velocity difference
v_a = velocity of sound in gas
v_d = root mean square thermal velocity of particle
v_e = particle electrical migration velocity
v_F = velocity fraction of cloud compared with single-particle velocity
v_g = velocity of gas or fluid
v_L = particle magnetic drift velocity
v_s = terminal settling velocity
v_∞ = velocity of medium at large distance from particle
V = volume
V_F = velocity fraction

x = distance
x = x coordinate
x_s = stopping distance
X_A = mole fraction of A in solid or liquid phase

y = system width
y = y coordinate
Y_A = mole fraction of A in vapor phase
Y_o = distance from stagnation to limiting streamlines
Y_{O_2} = mole fraction oxygen in gas phase

z = z coordinate

GREEK

α = coefficient of thermal reflection, see Eq. (4.45-4.46)
α = relative standard deviation
α_T = thermal accommodation coefficient, see Eq. (4.49)
β = divergence of gas stream leaving a jet
γ = ratio of specific heats
δ = drag
δ = partial derivative operator
δ = displacement distance
$\overline{\delta_d}$ = diffusion layer thickness
$\overline{\Delta x_d}$ = mean particle diffusional displacement, cm
∇ = del operator
ϵ = relative dielectric constant
ϵ = porosity
ϵ_i = individual fractional efficiency for removal of specific size particles
ϵ_o = dielectric constant of free space, equals 8.85×10^{-14} coulomb-cm/(cm^2-volt) or 8.85×10^{-12} coulomb^2/(joule-m)
θ = inclination angle or angle
λ = wavelength of light or wavelength
λ_g = mean free path of gas, equals 6.53×10^{-6} cm for air at SC
μ_g = viscosity of gas or fluid, equals 1.83×10^{-4} g/(cm-sec) **for air at SC**
μm = micrometers or microns
ρ_g = density of gas
ρ_p = density of particles
σ_{AB} = Waldmann-Schmitt factor, dimensionless
σ_g = standard geometric deviation

τ = characteristic time

τ = stress tensor

τ = relaxation time = $(d^2 \, \rho_p)/(18 \, \mu_g)$, sec

Υ = liquid surface tension

ϕ = fraction of volume occupied by particles

ϕ = phase angle or angle

χ = dynamic shape factor, dimensionless

ψ = sphericity factor, dimensionless

ψ = stream function

ω = angular velocity

ω = sound frequency

ENGLISH AND METRIC EQUIVALENTS

Length

    1 in. = 2.54 cm

    1 ft = 0.305 m

    1 micrometer = 10^{-6} m = 10^{-4} cm

Volume

    1 ft^3 = 0.0283 m^3 = 28.32 ℓ

    1 gal = 3.785 ℓ

Mass

    1 lb = 453.6 g

Pressure

    1 atm = 14.696 lb/in.^{2} = 76 cm Hg ($0°$C) = 407.2 in. H_2O

        = 760 torr = 1.01325 bar

    1 microbar = 1 dyne/cm^2

Speed

    1 ft/sec = 30.48 cm/sec

    1 ft/min = 0.508 cm/sec

Rate

    1 scfm = 1.6 Nm^3/hr

    1 gpm = 0.227 m^3/hr

Others

    100 ft^2/1000 cfm = 0.197 cm^2/(cm^3/sec)

    1 gal/1000 ft^3 = 0.134 ℓ/m^3

    1 grain/ft^3 = 2.29 g/m^3

    1 coulomb = 6.24 x 10^{18} electrons = 1 watt sec/volt

        = 10^7 g cm^2/(sec^2 volt)

    1 newton = 1 Kg m/sec^2 = 10^5 dynes

INDEX

acceleration 45,51
acoustic force 97-100
acoustic radiation pressure 98
adhesive force 115,141-144
adsorption 55,100,145,146
aerocolloidal systems 15
aerosol charge neutralizers 169
aerosol generation 172-179
agglomeration 144
air conditioning 76
air-to-cloth ratio 123
antinodes 97
aspect ratio 37
atomization
  constant feed 174
  pneumatic 116
Avagadro's number 93
axial velocity 42

Boltzmann's ideal gas relation 93
Brownian motion 30,60-62,73,74,
  96,101,102
bubbles
  formation 73
  rupture 138
buoyant force 25

cake-filter resistance 110
cascade impactor 155-166,174
  design 160
  efficiency 161
  Lundgren 193
  multi-jet 160
  operation 158

cellulose thimbles 186
cenospheres (fly ash particles) 10,37
centrifugal force 100
Chapman-Enskog Equation 140
characteristic times
  cleaning 119,134
  collection 115,123,131,134
  droplet life 120
  relaxation 32,33,35,45,46
  residence 115,116,125
  scrubbing life 120
charge 62,74-82,90,131
  droplet 133
  elementary 10,62
  image 79
  maximum 78
  removal time 133
  saturation 79,82
charged drop lifetime 133
charged filter bed 132
charging
  contact 74,75
  diffusional 75,83
  droplet 133
  electrolytic 75
  field strength 83
  gaseous ions 74
  induction 133
  time 79
cloud chambers 169
clouds 13,39-42
coalescence 144
collection, particle
  devices 111,115 ff.,125 ff.
  filter 121
  inertial 116

collection, particle
  sedimentation 123
  Also see impactors, electrostatic
    precipitators, filters and scrubbers
collection efficiency 71-74,89,116,
    123,147,165
  mass 123
collection time 119,133
  precipitator 131
  scrubber 119
collision atomizers 172
concentration determination 184
condensation 93,145
condensation aerosol 13
condensation nuclei 178
  counters 166,169
contact charging 75
contact separation 75
continuity, equation of 24
control devices, hybrid 131-135
corona 75,76
  back 111,115
  charging 75
coulombic force 79,86,89,90
Cunningham Factor 19,33,35,69,
    77,90
curvilinear motion 59-64

Darcy's Law 109,123
deceleration 45,51
density 10,24
departure tables 139
deposition 54,55,82-89,123
  eddy 127
  efficiency 54
  velocities 56
  Also see migration velocities
diameter 6
  aerodynamic 7,34-37,161
  cut 8,146,147
  equivalent 7
  mass mean (MMD) 3
  mean 3
  median 6
  projected 8
  sedimentation 7
dielectric constant of free space 131
dielectric constant of vacuum 79
diffuser elements 125

diffusion 10,30-35,56,91,92,96,97
  batteries 166-169
  charging 79
  displacement 33,35
  eddy 102-104
  equimolar counter (ECD) 96
  half-life 58
  stagnant (SD) 96
diffusive deposition 55-59
diffusivity 31,32,35
diffusiophoresis 31,91-93,96,100,
    101,140
dimensionless drag coefficient 27
dimensionless length 102
dimensionless numbers 16-20
dimensionless particle radius 102
dimensionless velocity 102
dispersion aerosol 6,13
displacement, mean 32-35
distance traveled by particles 50-52
drag force 25
drift velocities 98
  magnetic 90
drizzle 39
dust 14,15,122

eddy diffusion 102-104
electrical analyzers 171
electrical breakdown 115
electrical drift 115
electrical mobility 11,76,77
electrical velocities 85
electrofluidized beds 133
electropacked beds 133
electrophotophoresis 95
electrostatic augmentation 86,132
electrostatic force 74-89,141
electrostatic precipitation 85,111-
    115,125-131
  collection efficiency 87
  residence times 83
electrostatic resistivity 111
energy, equations of 25
equations, definitions 24-26
ESP
  See electrostatic precipitation
Eüler Equation 25
evaporation 93,145
explosiveness 12,14

fabric, napped and needle-punched 109
field charging 79,83
field strength 62,131
Fick's Law 31
filters 107-111,121-125
  efficiencies 122
  fabric 108,121-124
  fibers of 108,112
  inline 186
  pressure drop in 110
  pulse jet 123
  shaker 125
flow
  compressible 18
  hydrodynamic 91
  incompressible 18
  laminar 42,52,53
  potential 42,59
  separated-entrained 138
  viscous 42,59
flow meters 182
fluid
  incompressible 42
  Newtonian 24,42
fluorescence 169
fly ash 38,114
  size measurements 4
forest fires 39
Fourier Number 57,58
friction factor 25,27
Froude Number 19,65

gaseous ions 10,75,77
gas flow velocity 153
gas measurements 186
gas molecule mean free path 93
gas sampling 152
gas sneakage 125,127
Gauss Law 131
gravitational force 26-29
gravitophotophoresis 95
Greenburg-Smith impinger 69

Hi-Volume sampler 185
hydrodynamic flow 91
hydrodynamic interaction 39
hydrophilic particles 72

hydrophobic substances 72
hydrostatic force 39-41
hygroscopic particles 10

image forces 79,89
impaction 34,60,67,107
  data 164
  efficiency 63,70,71,117,123
  of ions 133
  parameters 19,107,116
impactors 116-121
  cascade 155-166,193
  jet velocity 160
impingers 69,181-184
inclusion, angle of 62
inertia 62,67-74,102
interception 60,63,64,107,123
internal circulation 37
internal energies 94
ions, gaseous 10,75,77
ion diffusion 75
ionization 75,76
irrigation 128
isokinetic sampling 152

kinematic coagulation 144
kinetic-molecular theory 17,30
Knudsen Number 17
Kozeny-Carman coefficient 110

Lagrangian frequency 62
laminar flow settling 52,53,58,85
light scattering 11
light, speed of 90
limiting-flow orifices 181,182
linear correlation 156
log-normal distribution 2,6
Lorentz magnetic force field 86, 89
louvers 137

Mach number 17
magnetic field intensity 90
magnetic force 89,90
magnetic particles 90
magnetophotophoresis 95

manometers   153,181
  inclined   183,184
  U-tube   183
Maxwell's Equation   145
mean path, apparent   30
mean free path   17
  apparent, particle   33,35
  gas molecule   93
mesh, knitted   137
midget impingers   69
migration velocity   86,88
mists   15,39
  elimination   136-138
  separators   136
mobility
  electrical   76,77
  mechanical (thermal, Brownian)
    31,33,35,76,77
molecular weight (gas)   154
monodispersity   2,6
motion, character of   41
motion, equation of   24

Navier-Stokes Equation   24
Nernst-Einstein Equation   31
Newton Drag Equation   27
Newton's Second Law of Motion   24
nodes   97
nucleation   145
numerical aperture (NA)   11

Ohm's Law   111
optical counters   169,170
optical effects on particle movement
    11
orifice scrubbers   118,181,182
origination states   13-15
oscillating force   61-63

particles, primary and secondary 13
particle properties   10-12
Peclet number   19,74
penetration   116,122
permeability   109
phoretic forces   72,90-95,121
photophoresis   91,95
piezoelectric microbalance sensor
    171

pitot tubes   153,186
  standard   152
  Stauscheibe   187
  Type S   187
plate length, critical   53
polar force   141
polydispersity   2,6
pore size   109
porosity   110
Prandtl number   18
precipitation velocity   86
precipitator
  space charge   132
  See also electrostatic precipitation
pressure   138-140,189
pressure drop   116,118
pressure force   24
probe   152
pulse jet filter   123

quasistationary motion   50
quiet zone   102,104

radiometric forces   91-95
rain   39
rapping   115,125,128
reentrainment   107,108,115,128-
    130,138
  velocity   136,137
relaxation time   32,35,45,46
residence time   115
resistivity   111,114
resonance   97
Reynolds Number   9,16,17,49,52,
    54,65
Ringelmann chart   185
root mean square velocity
  thermal   32-35
  turbulent   104

saltation   115
sampling   151,152
  error categories   154
  gas   152
  isokinetic   152
saturation impaction charges   79,
    82,134
saturation vapor pressure   92

scale-up of systems   64,65
scavenging   72-74
Schmidt Number   18
scrubbers
  charged drop   132
  cleaning time   119
  orifice   118,181,182
  spray   120
  steam hydro   132
  tests on   121
scrubbing   72,73
  life time   120
sedimentation   123
self-agglomerator   132
separator
  efficiency   137
  mist elimination   137
  pressure drop   137
  wire mesh   137
settling   52-54,123
  efficiency   53
  turbulent flow   54
  velocity, cloud   42
shape factor   7
  dynamic   8,9
  effects on particles   37-39
sieve plates   73
sieving   108
size distribution   2-6,164
  cumulative   165
size measurement   151-179
size regimes   20,21
  continuum   20,26-28,73
  Cunningham   20
  Free Molecule   29,92,96
  Slip Flow   19,20,28,29,73,97
  Transition   20,92
sizing techniques   151,168
smog and smoke   15
sonic velocity   88
Soret effect   30
Souders-Brown Equation   137
specific heats   18
specific surface   11
sphericity factor   8,9
spray electrification   75
standard deviation
  geometric   2
  relative   6
static pressure   189

Stephan flow   91-93,100,138,140
Stokes Law   25,37,47,90
Stokes Number   18,63,65,69,107
Stokes-Einstein Equation   32
stopping distance   51,52
straining   122
stream function   52
streamlines   52,60,67,68,72
Student's "t" test for small samples
     155,156
surfaces   11-13,42
  compaction   113
  effects on particles   12,42,43
  resistivity   111
  variations   11
surface tension   11,142,176

temperature   138-140
terminal states   13-15
test equipment   181-184
thermal accommodation coefficient
     94
thermal diffusion   96
thermophoresis   90,93,94,96,101
thermophoretic force   93,94
trajectory   62
translational energies   94
tribo electrification   75
turbulent boundary layer   104
turbulent flow settling   54

ultrafine particles   178,179
ultrasonics   97,144
universal gas constant   17

van der Waals forces   142
vaporization, heat of   11
vapor pressure   11
velocity   46-50,188
  acoustic   17
  bulk   42,53
  distribution   52
  electrical   85
  filter gas   122
  gas flow   153
  maximum   42
  mean diffusional   33

velocity
  mean molecular 30
  mean square 32
  migration 86,88
  nonsteady-state 48
  profiles 60
  reentrainment 136,137
  root mean square 33,35
  stopping 48
  terminal settling 7,26,36,56,88
  thermal 33,35
venturi collectors 116

vibrating-orifice aerosol generator
  176
viscosity 139
viscous force 24
visual emissions 185
volume compaction 113

wavelength, light 11
wettability 116,118

zigzag baffles 137